EL LIBRO PERTENECE
A

TABLA DE CONTENIDO

POSTURAS DE YOGA PARA PRINCIPIANTES

TABLA DE CONTENIDO
POSTURAS DE YOGA PARA INTERMEDIOS

TABLA DE CONTENIDO
POSTURAS DE YOGA PARA EXPERTOS

POSTURAS DE YOGA PARA PRINCIPIANTES

1. POSTURA DE MONTAÑA

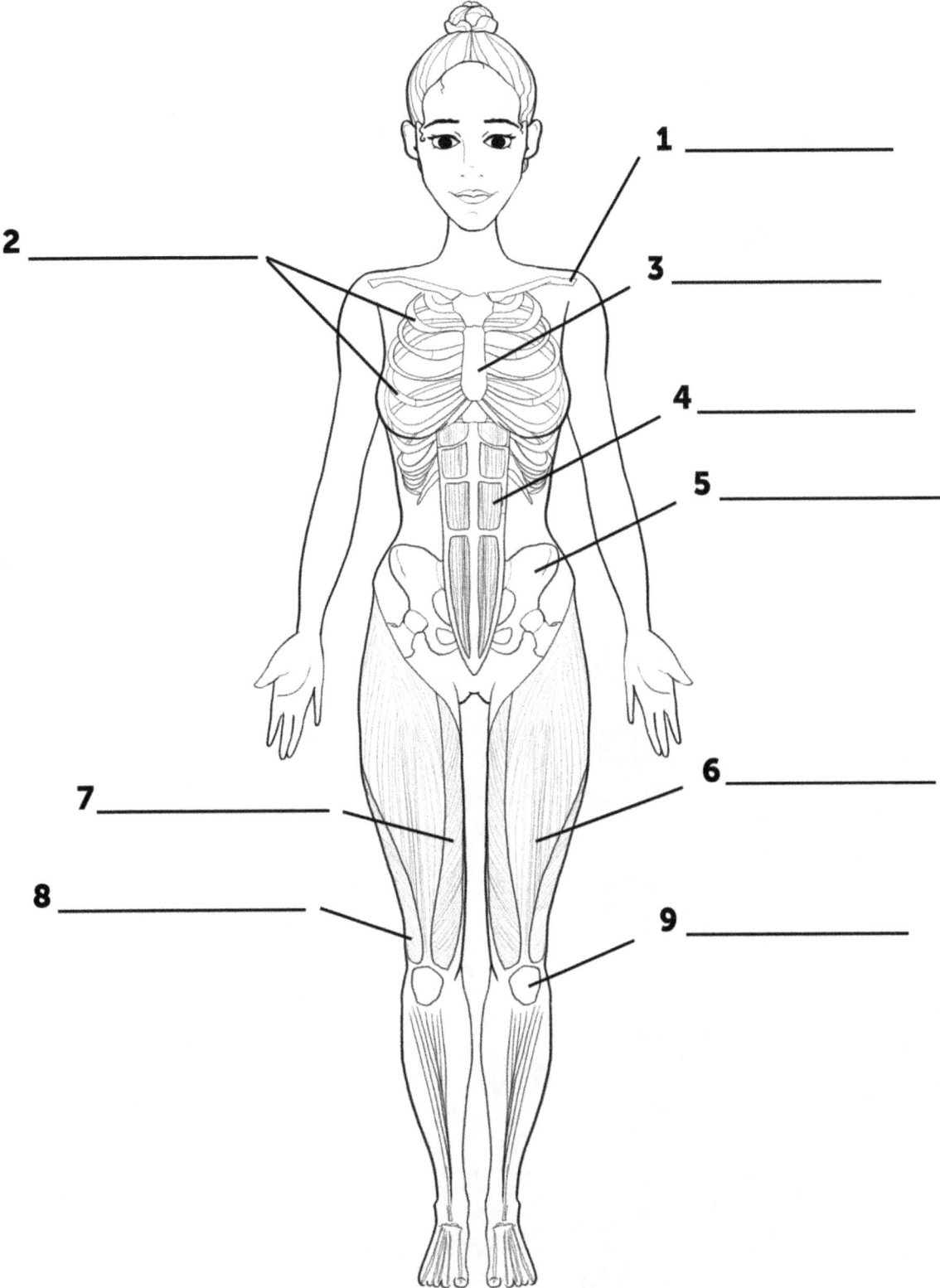

1 _____

2 _____

3 _____

4 _____

5 _____

6 _____

7 _____

8 _____

9 _____

1. POSTURA DE MONTAÑA

1. CLAVÍCULA
2. COSTILLAS
3. ESTERNÓN
4. RECTO ABDOMINAL
5. PELVIS
6. CUADRÍCEPS
7. VASTO MEDIAL
8. MÚSCULO VASTO LATERAL
9. RÓTULA

2. POSTURA DE LA PALMERA

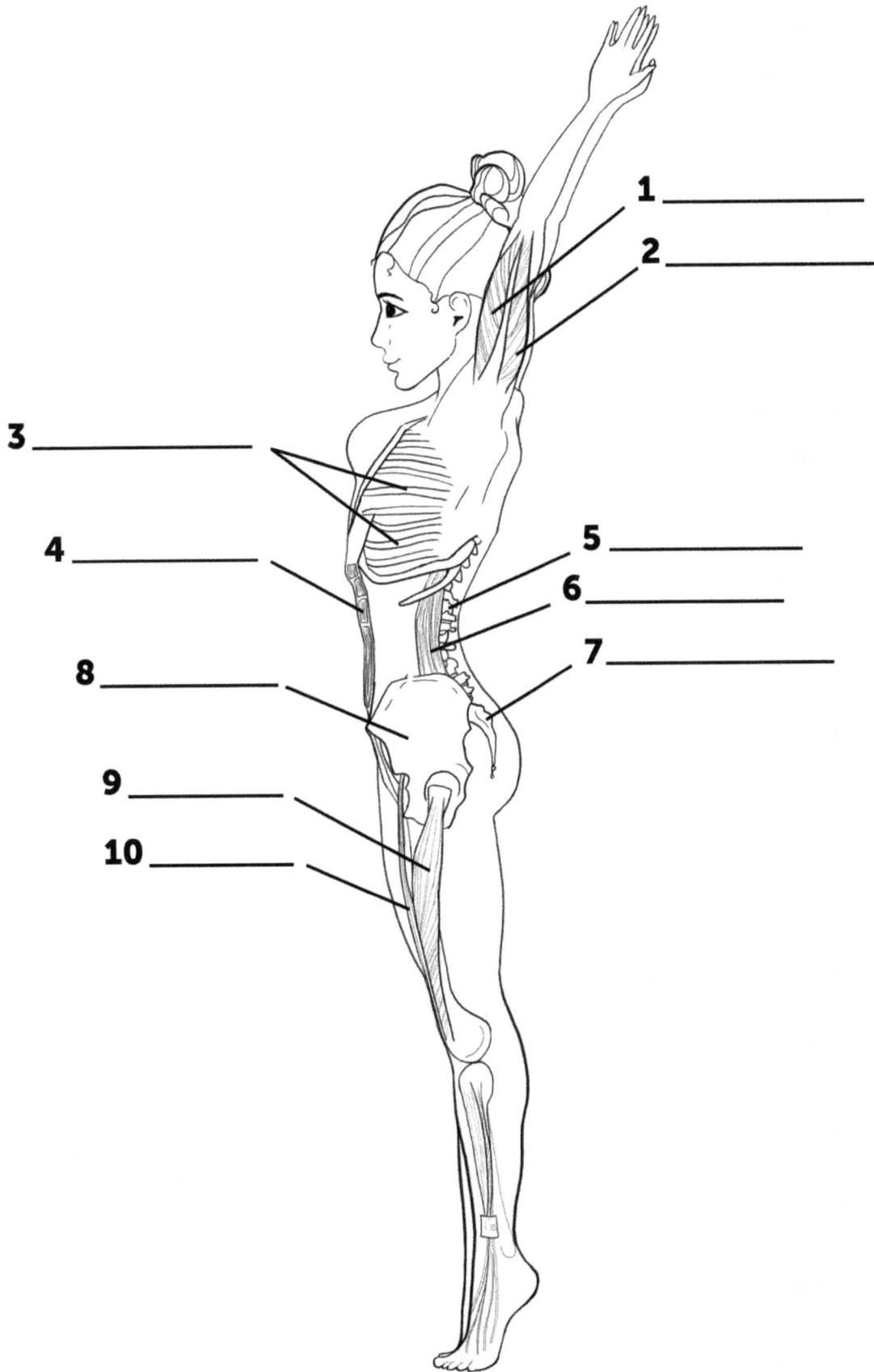

1 _____

2 _____

3 _____

4 _____

5 _____

6 _____

7 _____

8 _____

9 _____

10 _____

2. POSTURA DE LA PALMERA

1. TRÍCEPS BRAQUIAL

2. DELTOIDES

3. COSTILLAS

4. RECTO ABDOMINAL

5. COLUMNA VERTEBRAL

6. ERECTOR DE LA COLUMNA

7. SACRO

8. PELVIS

9. RECTO FEMORAL

10. SARTORIO

3. URDHVA MUKHA SHVANASANA

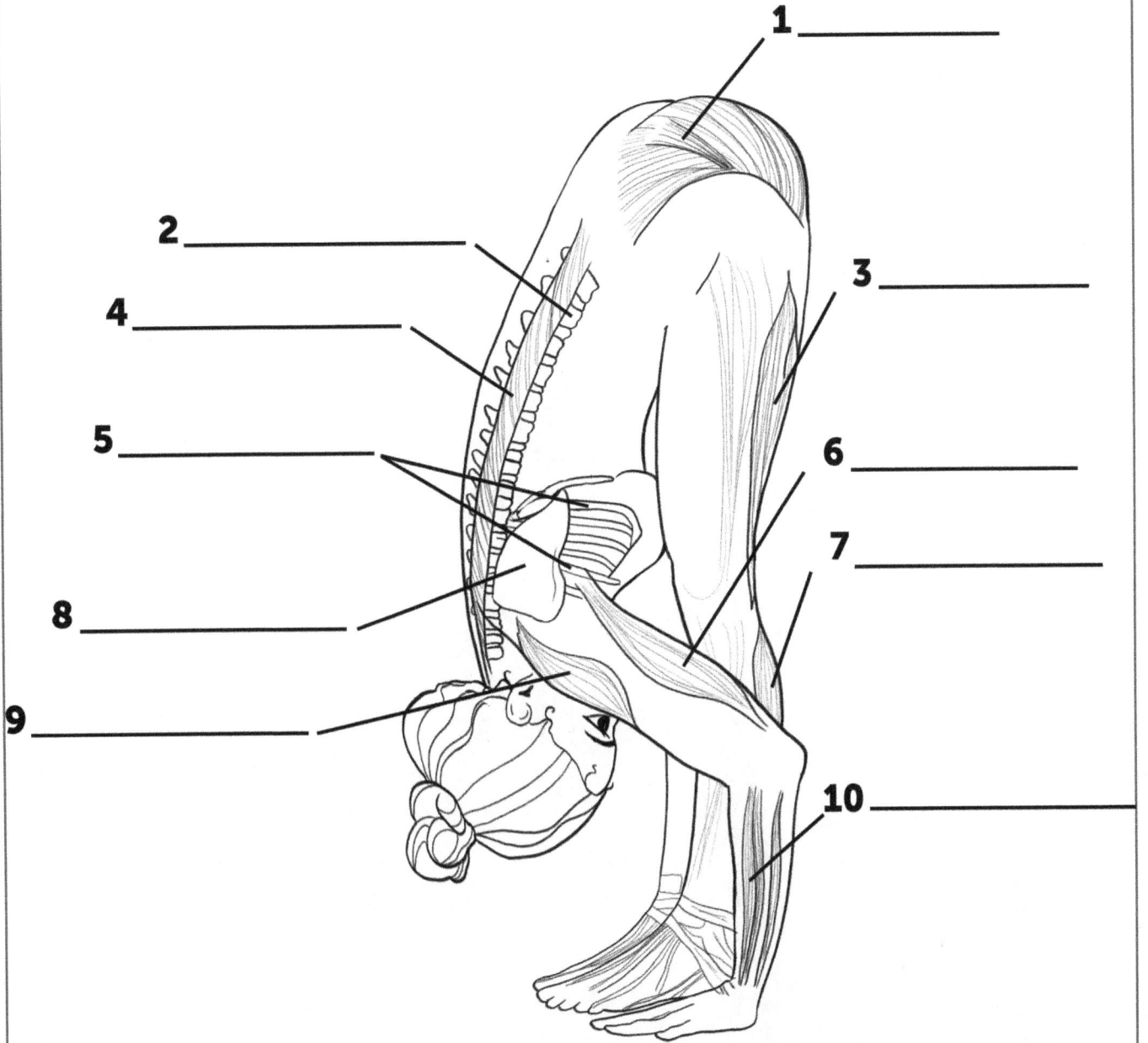

1 _____

2 _____

3 _____

4 _____

5 _____

6 _____

7 _____

8 _____

9 _____

10 _____

3. URDHVA MUKHA SHVANASANA

1. PIRIFORME
2. COLUMNA VERTEBRAL
3. ISQUIOTIBIALES
4. MÚSCULOS ESPINALES
5. COSTILLAS
6. TRÍCEPS BRAQUIAL
7. GASTROCNEMIO
8. ESCÁPULA
9. DELTOIDES
10. EXTENSOR DIGITORUM

4. ARDHA UTTANASANA

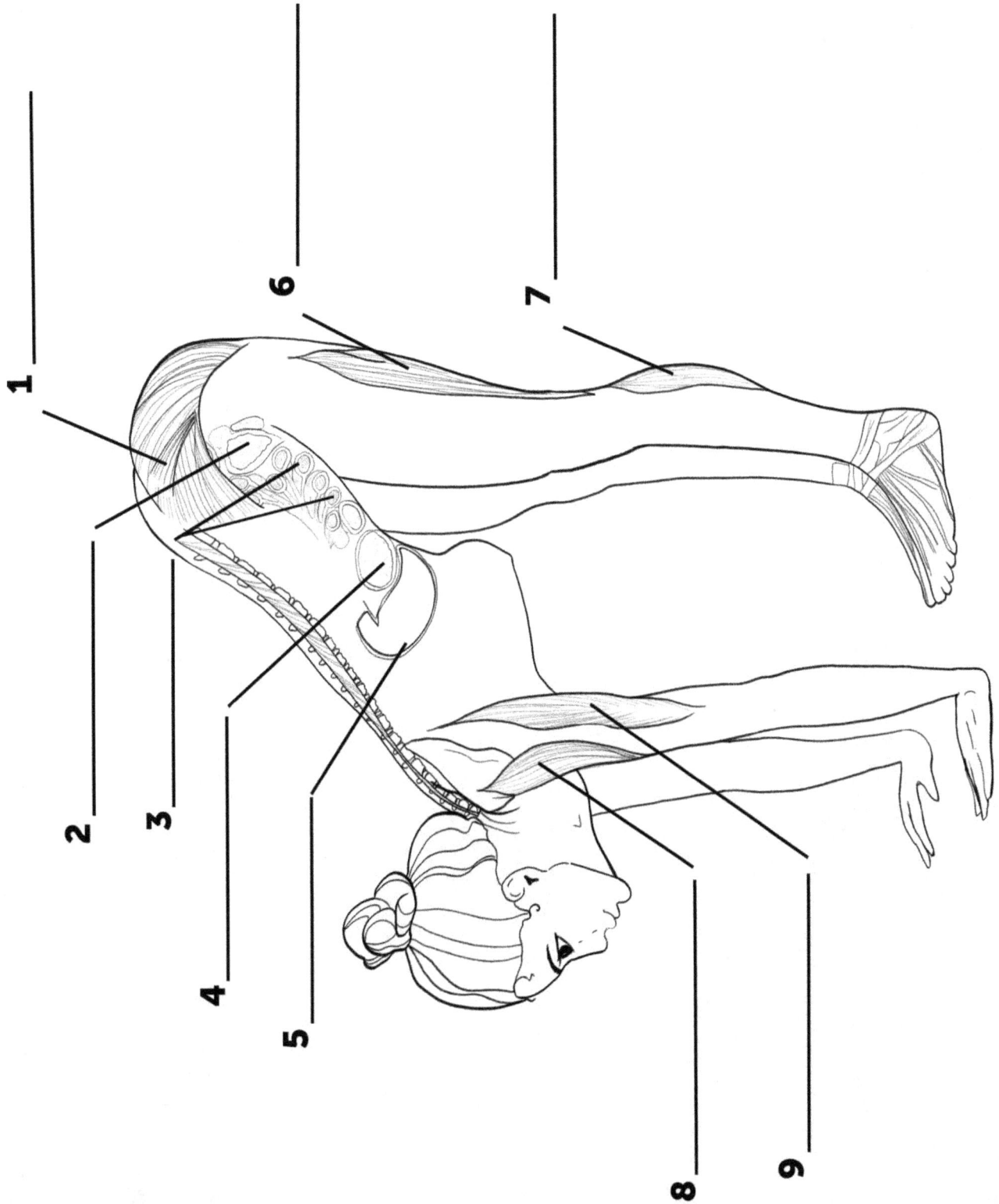

1

2

3

4

5

6

7

8

9

4. ARDHA UTTANASANA

1. PIRIFORME
2. VEJIGA URINARIA
3. INTESTINO DELGADO
4. ESTÓMAGO
5. HÍGADO
6. ISQUIOTIBIALES
7. GASTROCNEMIO
8. DELTOIDES
9. TRÍCEPS BRAQUIAL

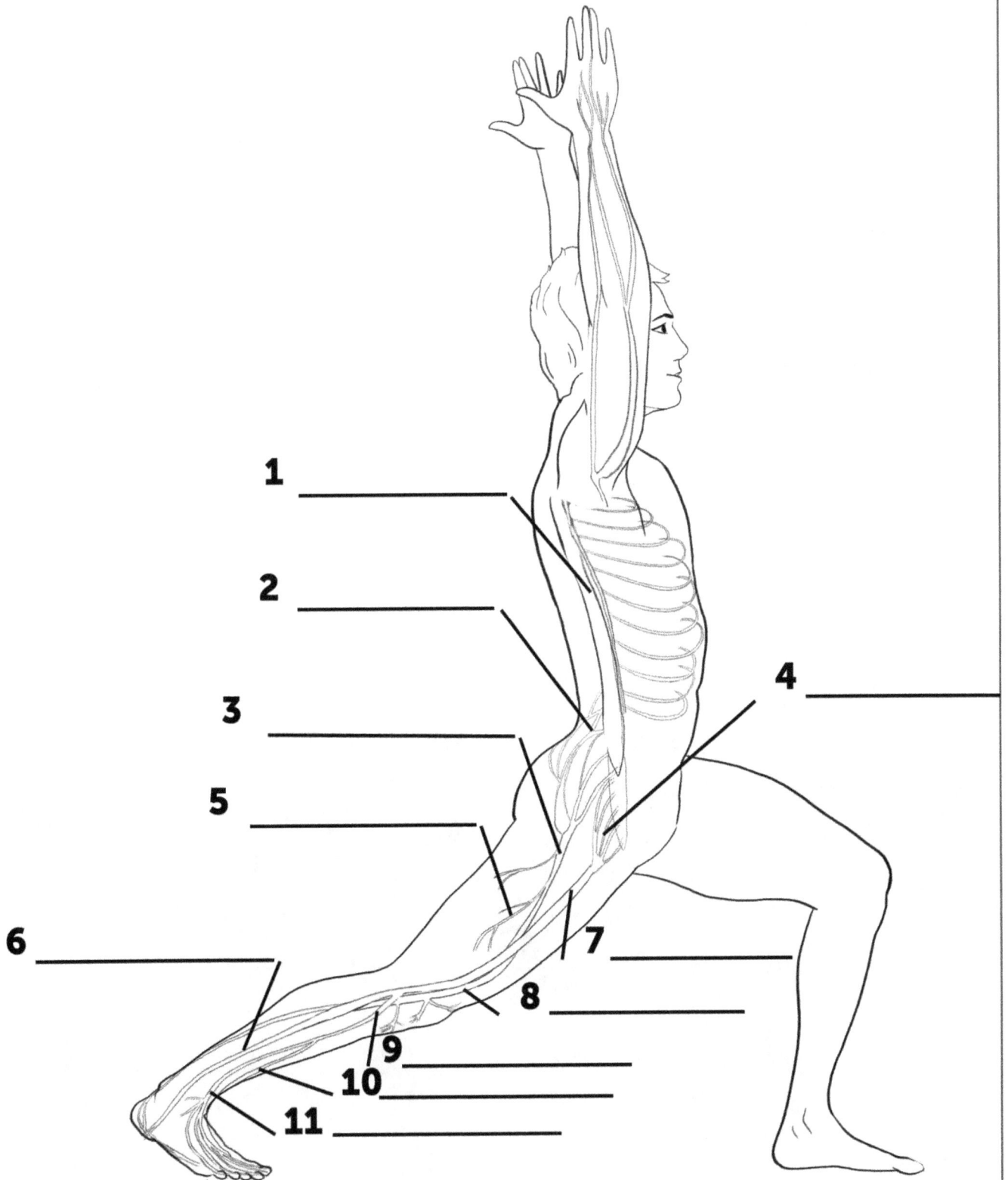

5. LUNGE ALTO

1 _____

2 _____

3 _____

4 _____

5 _____

6 _____

7 _____

8 _____

9 _____

10 _____

11 _____

5. LUNGE ALTO

1. MÉDULA ESPINAL
2. PLEXO LUMBAR
3. FEMORAL
4. PLEXO SACRO
5. RAMAS MUSCULARES DE FEMORAL
6. CIÁTICO
7. CIÁTICO
8. SAFENA
9. PERONEO COMÚN
10. SURAL
11. PERONEO SUPERFICIAL

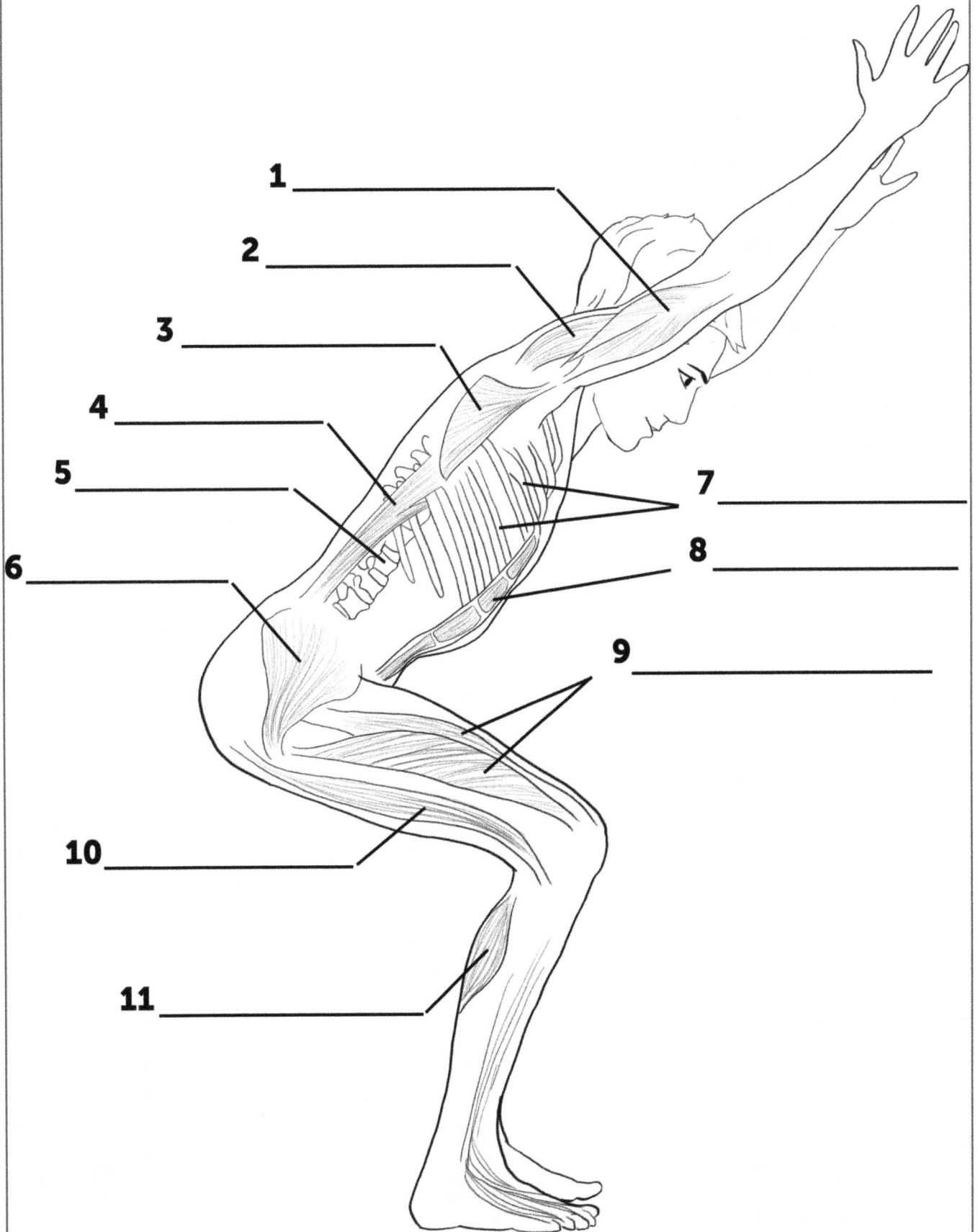

6. UTKATASANA

1 _____

2 _____

3 _____

4 _____

5 _____

6 _____

7 _____

8 _____

9 _____

10 _____

11 _____

6. UTKATASANA

1. TRÍCEPS BRAQUIAL
2. DELTOIDES
3. INFRAESPINOSO
4. ERECTOR DE LA COLUMNA
5. COLUMNA VERTEBRAL
6. GLÚTEO MEDIO
7. COSTILLAS
8. RECTO ABDOMINAL
9. CUADRÍCEPS
10. ISQUIOTIBIALES
11. GASTROCNEMIO

7. TRIKONASANA

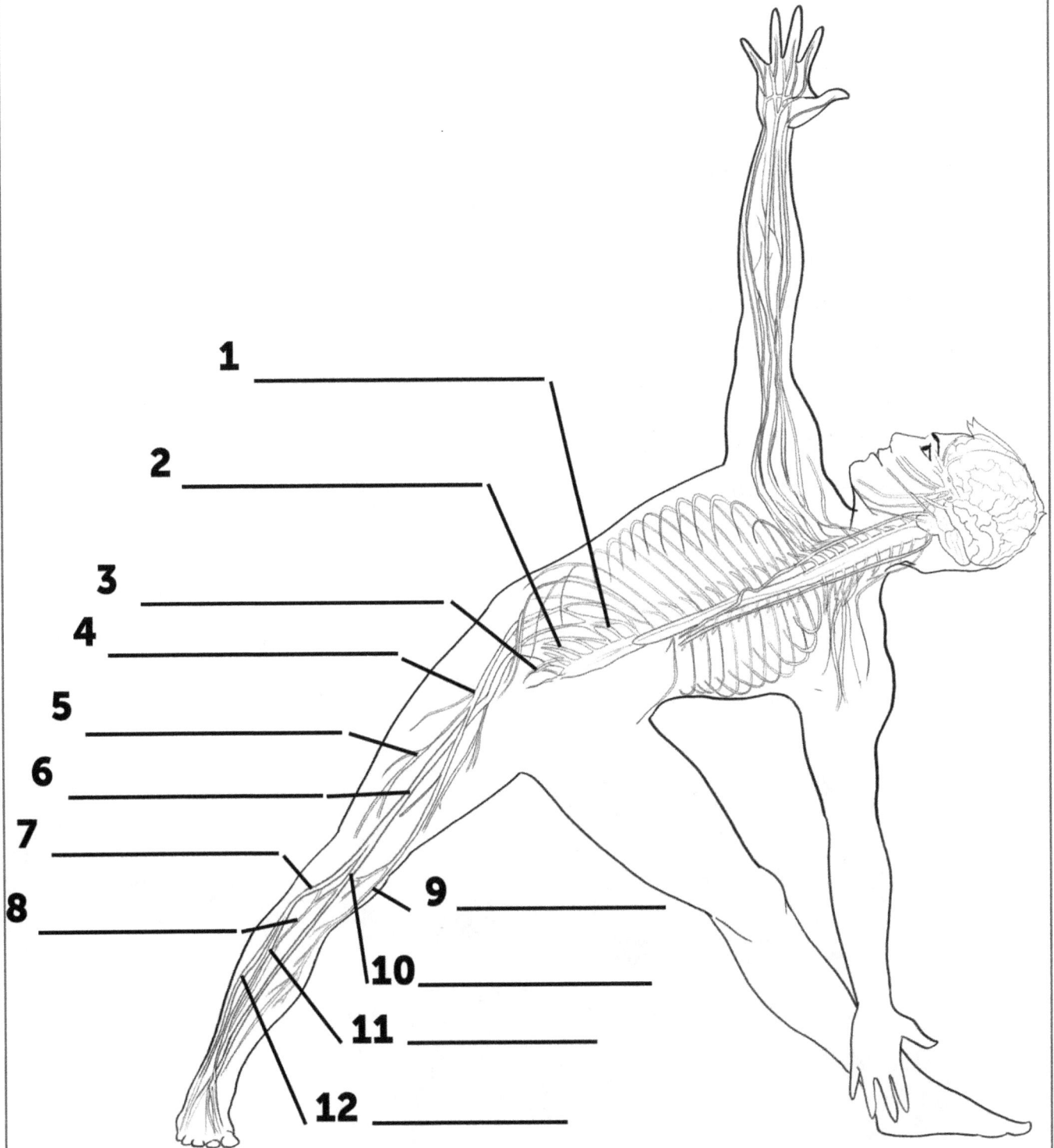

1 _____

2 _____

3 _____

4 _____

5 _____

6 _____

7 _____

8 _____

9 _____

10 _____

11 _____

12 _____

7. TRIKONASANA

1.   PLEXO LUMBAR

2.   PLEXO SACRO

3.   NERVIO PUDENDO

4.   FEMORAL

5.   RAMAS MUSCULARES DE FEMORAL

6.   CIÁTICO

7.   PERONEO COMÚN

8.   SURAL

9.   SAFENA

10.   TIBIAL

11.   PERONEO PROFUNDO

12.   PERONEO SUPERFICIAL

8. EUTTHITA PARSVAKONASANA

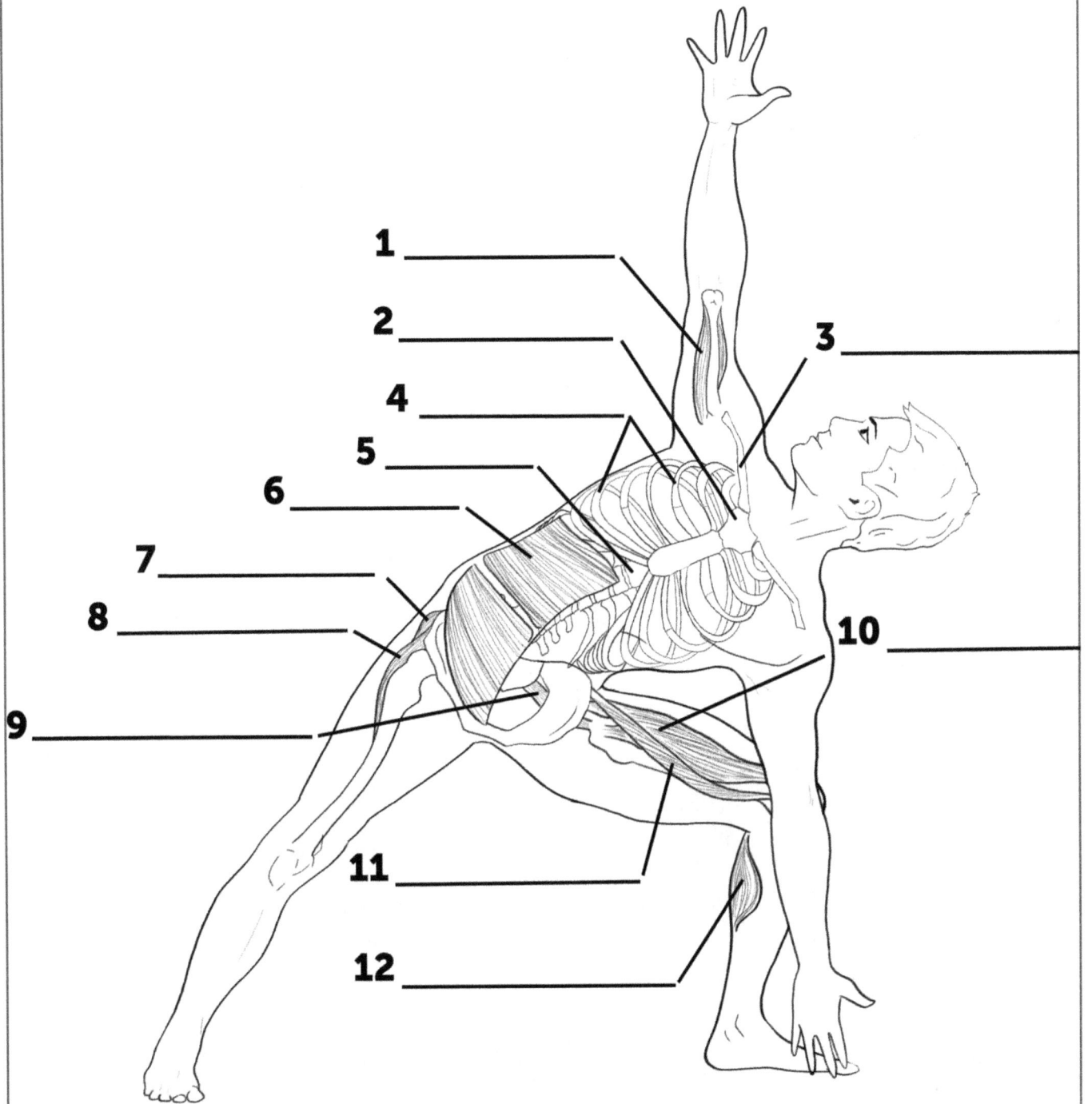

1 _____

2 _____

3 _____

4 _____

5 _____

6 _____

7 _____

8 _____

9 _____

10 _____

11 _____

12 _____

8. EUTTHITA PARSVAKONASANA

1. BÍCEPS BRAQUIAL
2. ESTERNÓN
3. CLAVÍCULA
4. COSTILLAS
5. COLUMNA VERTEBRAL
6. OBLICUO INTERNO
7. GLÚTEO MEDIO
8. MÚSCULO TENSOR DE LA FASCIA LATA
9. PIRIFORME
10. CUADRÍCEPS
11. SARTORIO
12. GASTROCNEMIO

9. DANDASANA

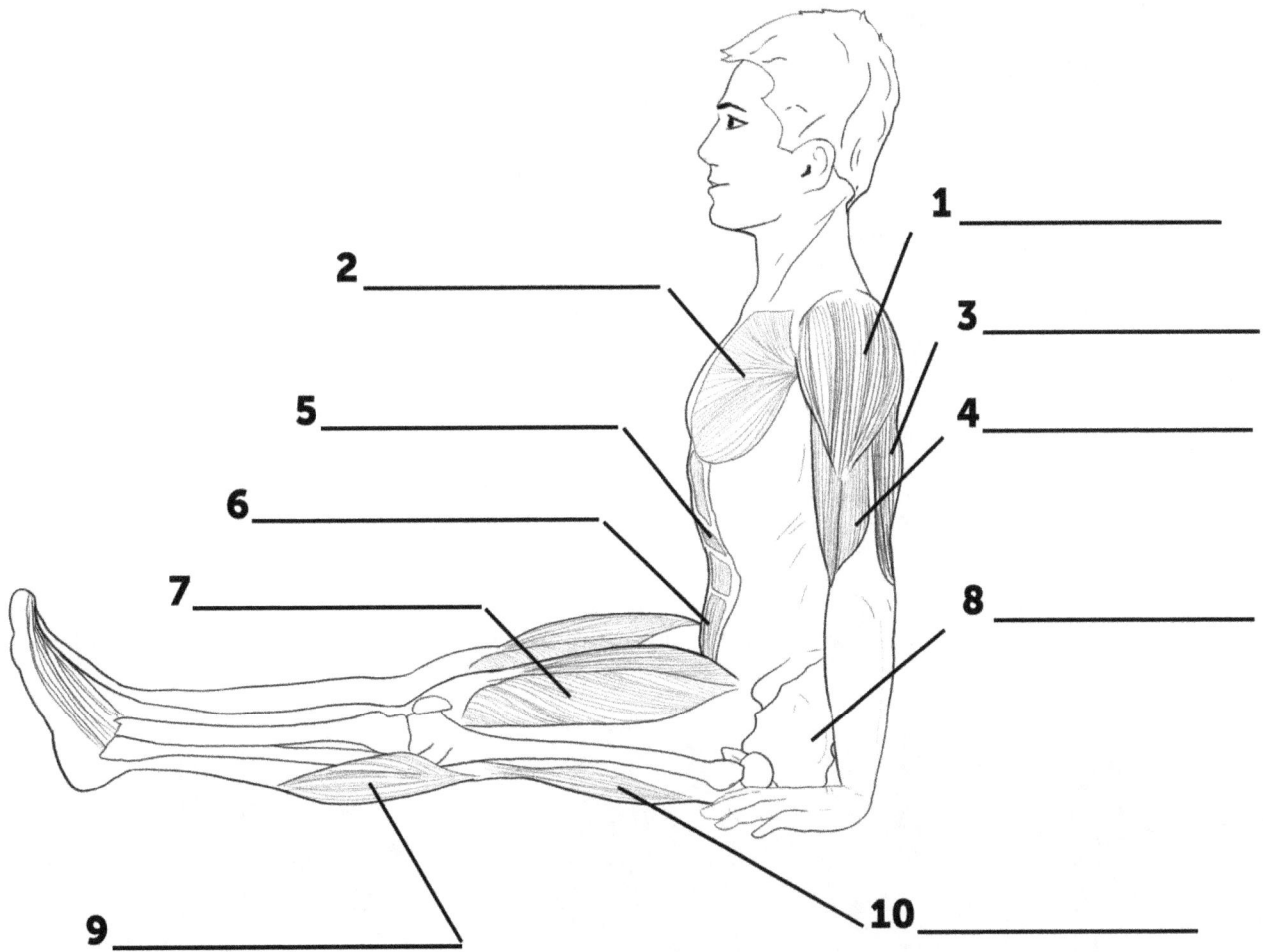

2 _____

5 _____

6 _____

7 _____

9 _____

1 _____

3 _____

4 _____

8 _____

10 _____

9. DANDASANA

1. DELTOIDES
2. PECTORAL MAYOR
3. TRÍCEPS BRAQUIAL
4. BÍCEPS BRAQUIAL
5. RECTO ABDOMINAL
6. MÚSCULOS DEL ABDOMEN INFERIOR
7. CUADRÍCEPS
8. PELVIS
9. GASTROCNEMIO
10. ISQUIOTIBIALES

10. SUKHASANA

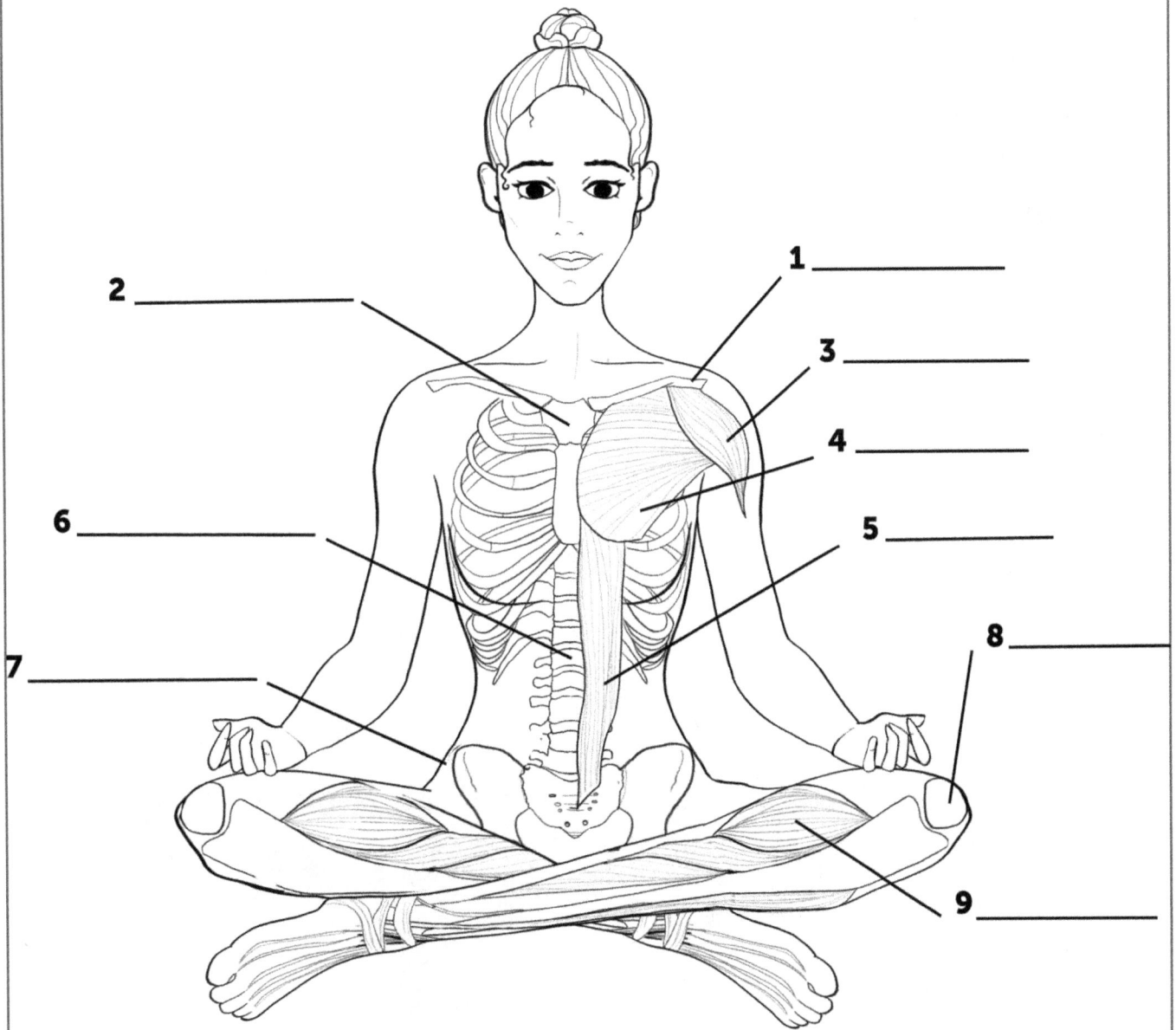

2 _____

1 _____

3 _____

4 _____

6 _____

5 _____

8 _____

7 _____

9 _____

10. SUKHASANA

1. CLAVÍCULA

2. ESTERNÓN

3. DELTOIDES

4. PECTORAL MAYOR

5. RECTO ABDOMINAL

6. COLUMNA VERTEBRAL

7. PELVIS

8. RÓTULA

9. GASTROCNEMIO

11. BADDHA KONASANA

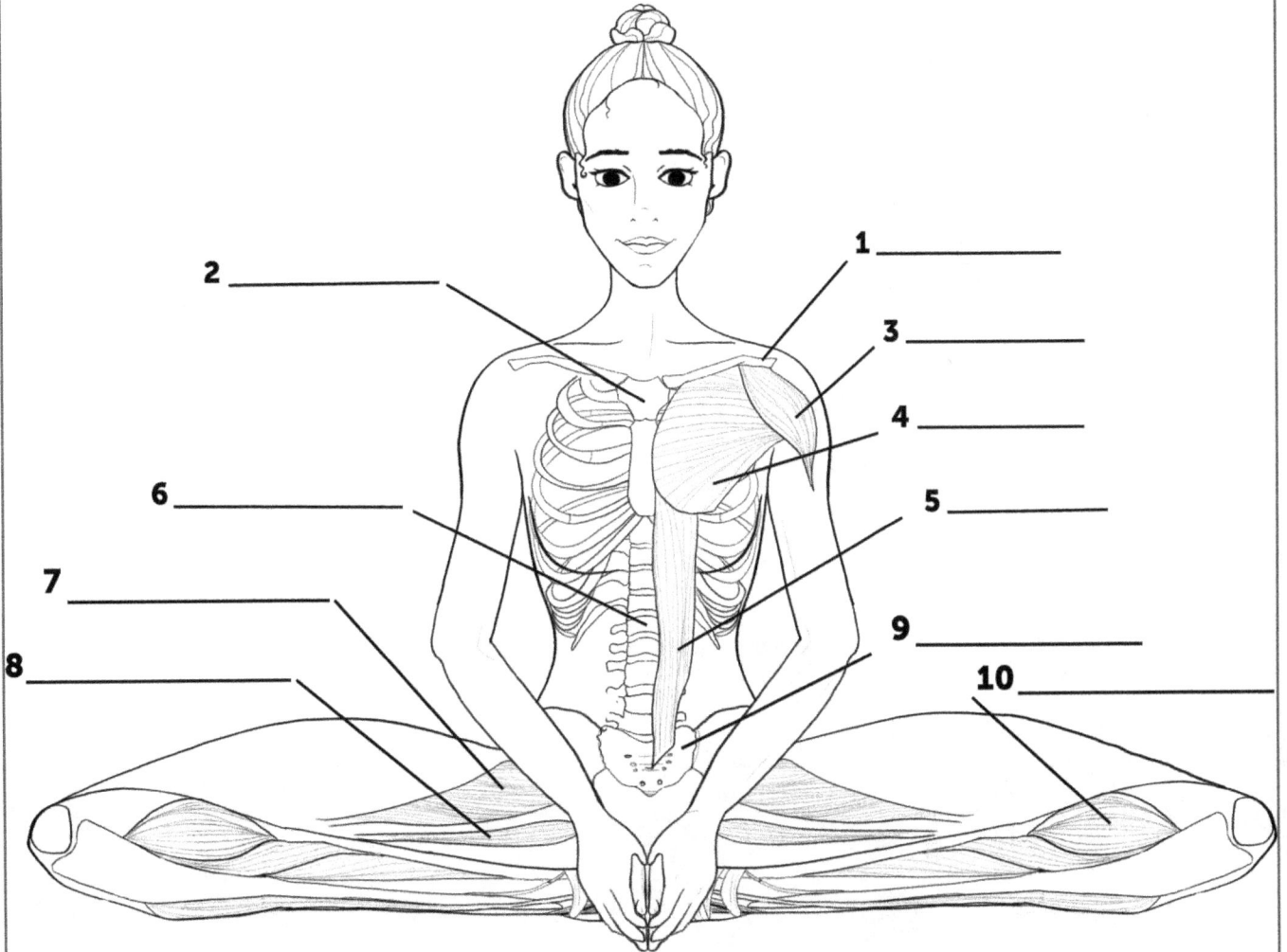

1 _____

2 _____

3 _____

4 _____

5 _____

6 _____

7 _____

8 _____

9 _____

10 _____

11. BADDHA KONASANA

1. CLAVÍCULA
2. ESTERNÓN
3. DELTOIDES
4. PECTORAL MAYOR
5. RECTO ABDOMINAL
6. COLUMNA VERTEBRAL
7. MÚSCULO ADUCTOR LARGO DEL MUSLO
8. GRÁCIL
9. SACRO
10. GASTROCNEMIO

12. ARDHA MATSYENDRASANA

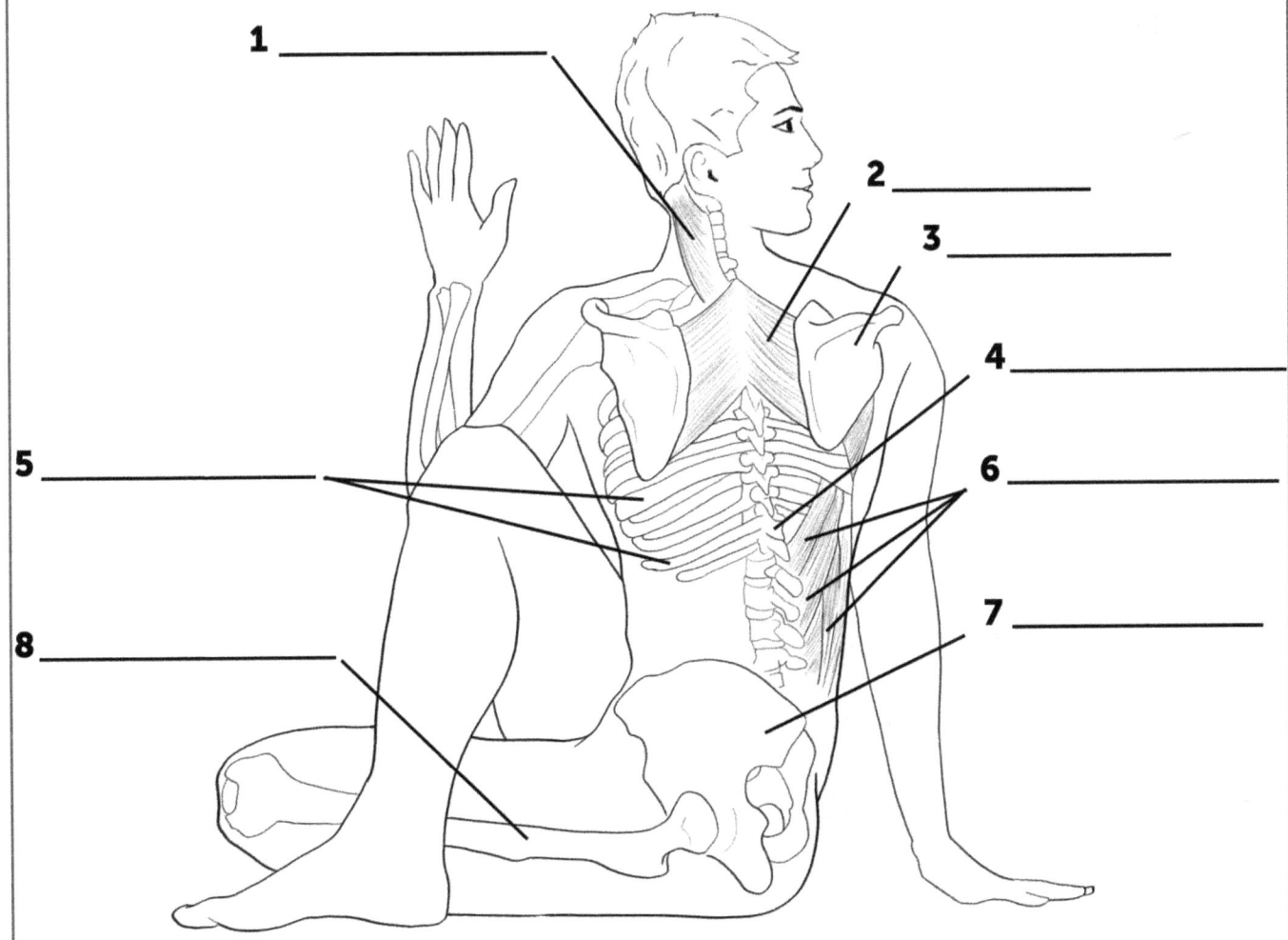

1 _____

2 _____

3 _____

4 _____

5 _____

6 _____

7 _____

8 _____

12. ARDHA MATSYENDRASANA

1.   MÚSCULO ESPLENIO DE LA CABEZA

2.   ROMBOIDES

3.   ESCÁPULA

4.   COLUMNA VERTEBRAL

5.   COSTILLAS

6.   ERECTOR DE LA COLUMNA

7.   PELVIS

8.   FÉMUR

13. POSTURA DE LA MESA

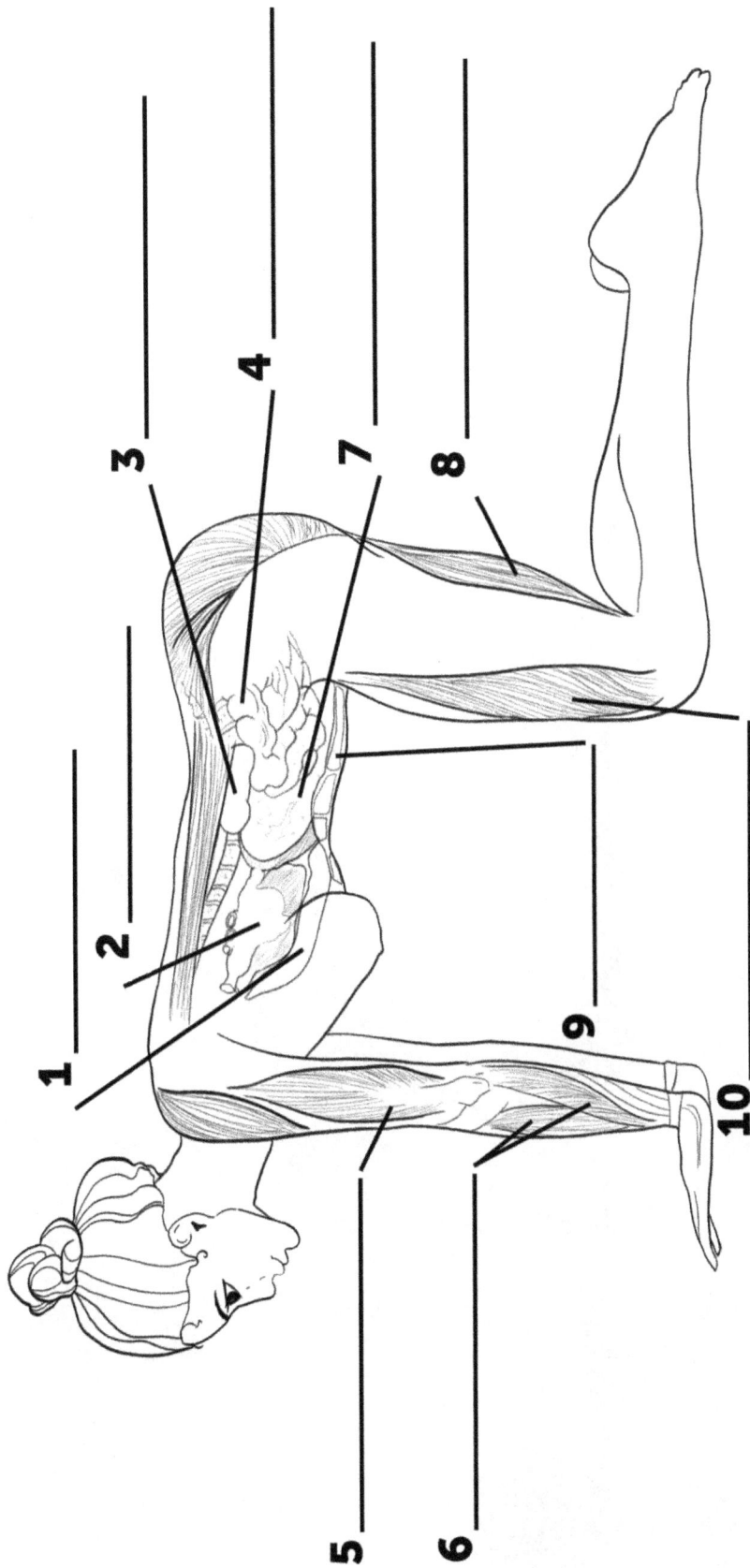

13. POSTURA DE LA MESA

1. PULMONES

2. CORAZÓN

3. RIÑÓN

4. COLON ASCENDENTE

5. TRÍCEPS BRAQUIAL

6. PRONADORES

7. HÍGADO

8. ISQUIOTIBIALES

9. RECTO ABDOMINAL

10. CUADRÍCEPS

14. POSTURA DEL GATO

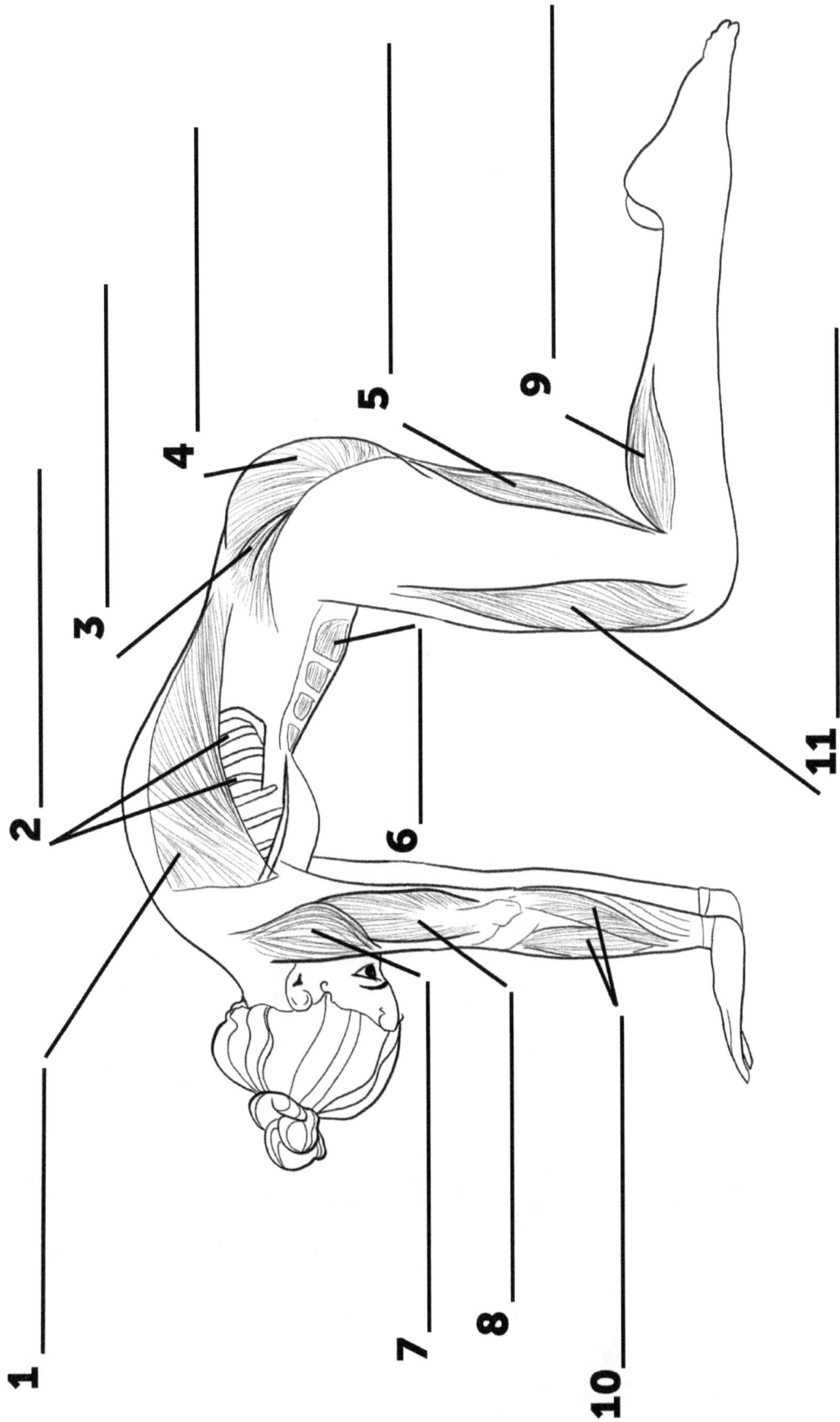

1

2

3

4

5

6

7

8

9

10

11

14. POSTURA DEL GATO

1. LATISSIMUS DORSI

2. COSTILLAS

3. PIRIFORME

4. MÚSCULO GLÚTEO MAYOR

5. ISQUIOTIBIALES

6. RECTO ABDOMINAL

7. DELTOIDES

8. TRÍCEPS BRAQUIAL

9. GASTROCNEMIO

10. PRONADORES

11. CUADRÍCEPS

15. POSTURA DE LA VACA

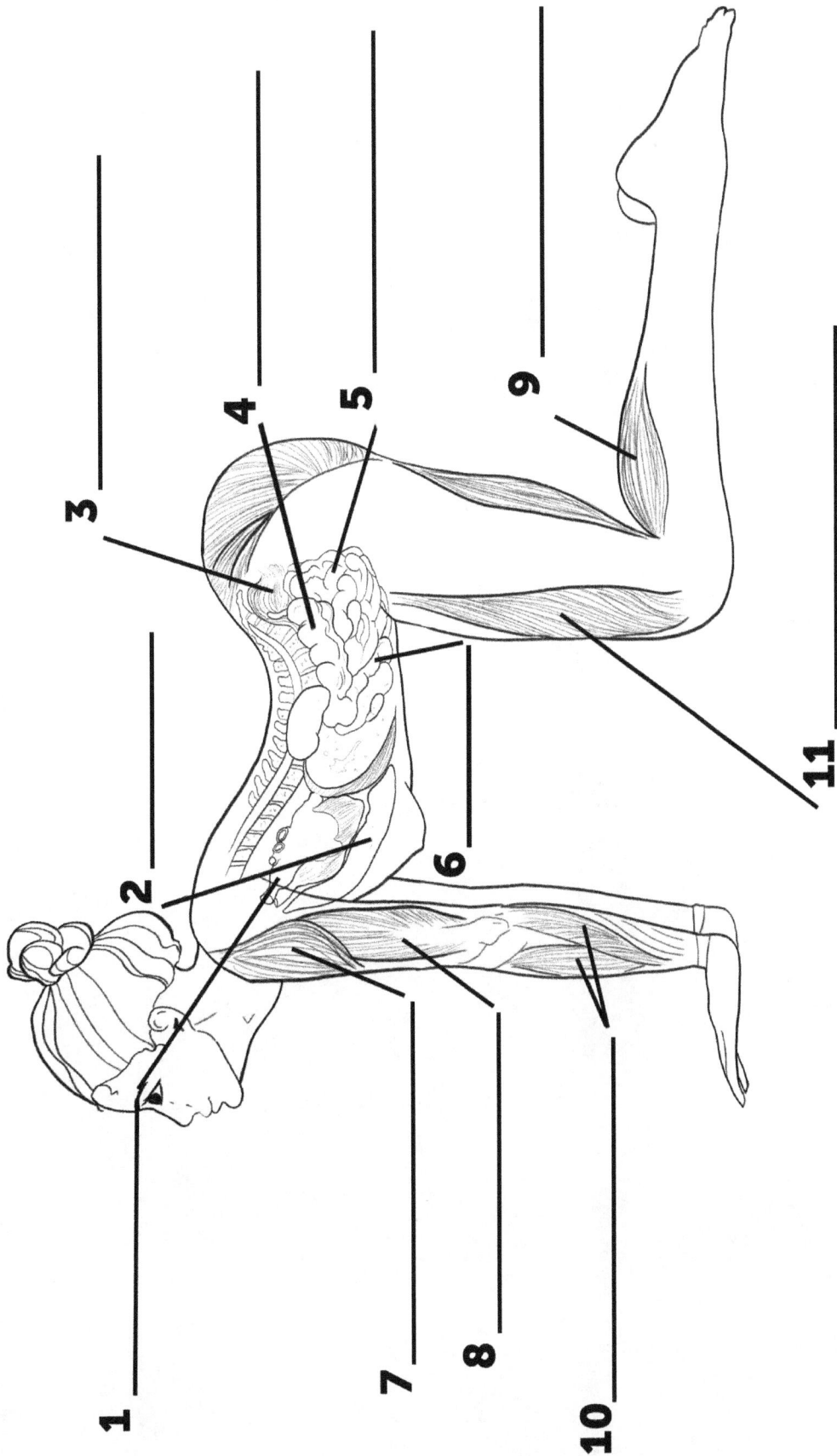

15. POSTURA DE LA VACA

1. CORAZÓN
2. PULMONES
3. RECTO
4. COLON ASCENDENTE
5. FOLICULOS DE INTESTINO DELGADO
6. COLON TRANSVERSO
7. DELTOIDES
8. TRÍCEPS BRAQUIAL
9. GASTROCNEMIO
10. PRONADORES
11. CUADRÍCEPS

16. POSTURA DE LA MESA DE EQUILIBRIO

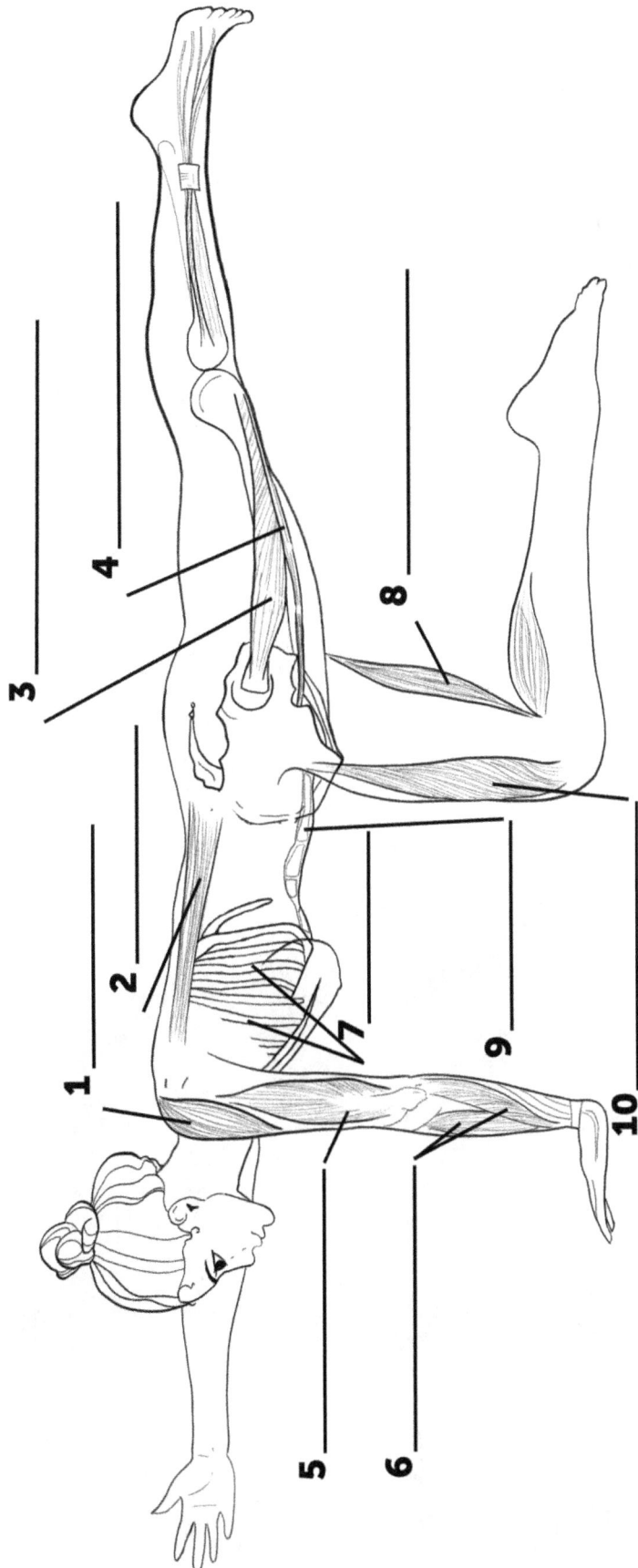

16. POSTURA DE LA MESA DE EQUILIBRIO

1. DELTOIDES
2. ERECTOR DE LA COLUMNA
3. RECTO FEMORAL
4. SARTORIO
5. TRÍCEPS BRAQUIAL
6. PRONADORES
7. COSTILLAS
8. ISQUIOTIBIALES
9. RECTO ABDOMINAL
10. CUADRÍCEPS

17. ARDHA PURVOTTANASANA

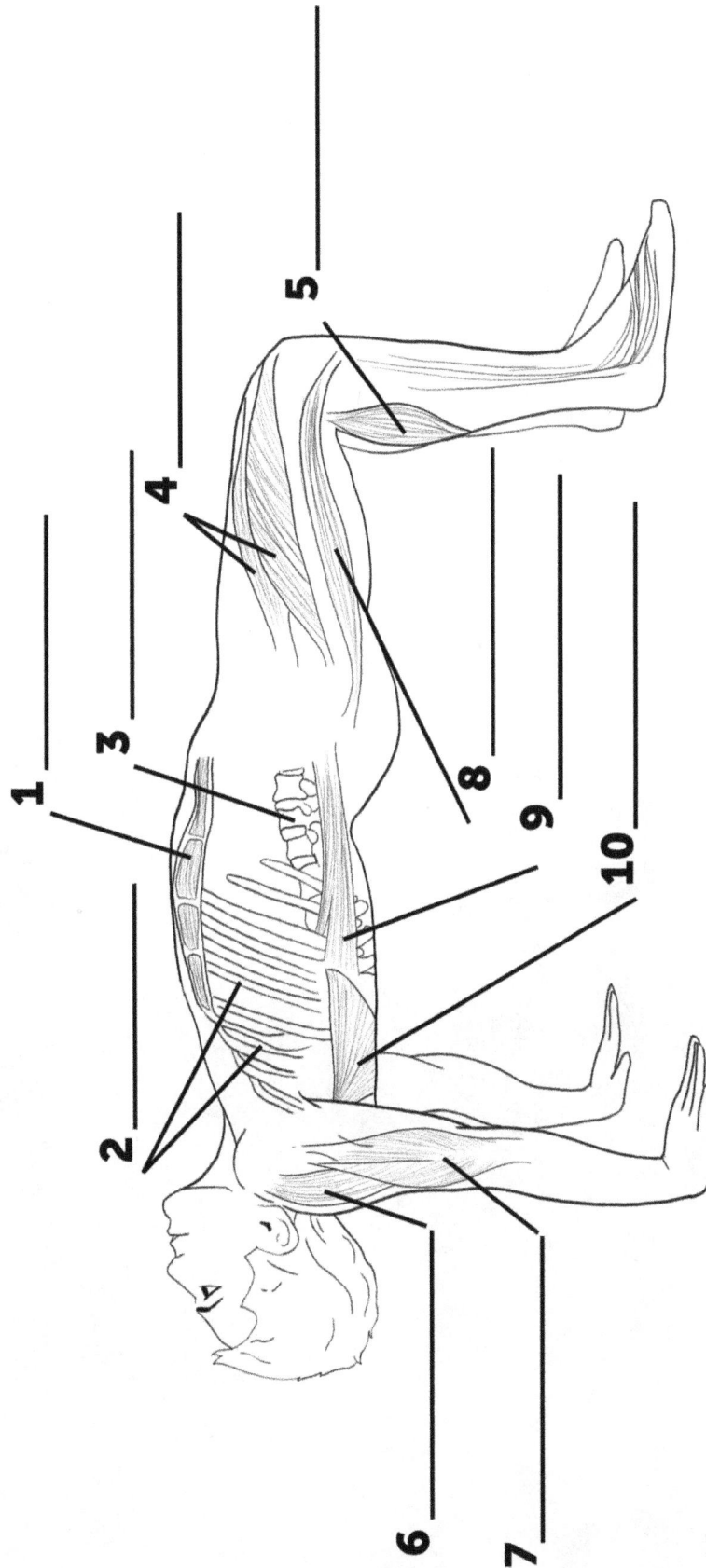

17. ARDHA PURVOTTANASANA

1. RECTO ABDOMINAL
2. COSTILLAS
3. COLUMNA VERTEBRAL
4. CUADRÍCEPS
5. GASTROCNEMIO
6. DELTOIDES
7. TRÍCEPS BRAQUIAL
8. ISQUIOTIBIALES
9. ERECTOR DE LA COLUMNA
10. INFRAESPINOSO

18. POSTURA DE LA ESFINGE

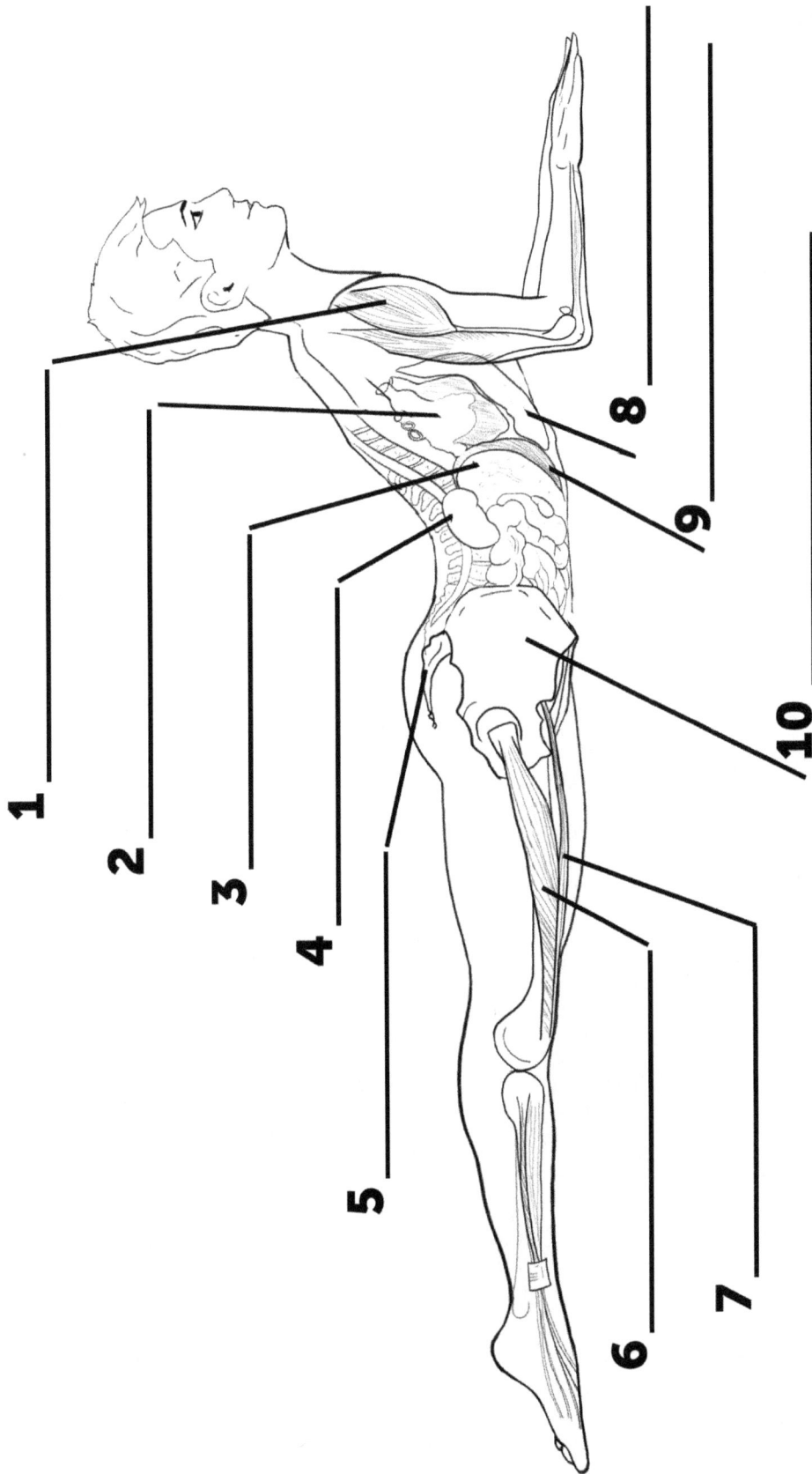

1

2

3

4

5

6

7

8

9

10

18. POSTURA DE LA ESFINGE

1.   DELTOIDES

2.   CORAZÓN

3.   HÍGADO

4.   RIÑÓN

5.   SACRO

6.   RECTO FEMORAL

7.   SARTORIO

8.   PULMONES

9.   DIAFRAGMA

10.   PELVIS

19. POSTURA DE LA COBRA

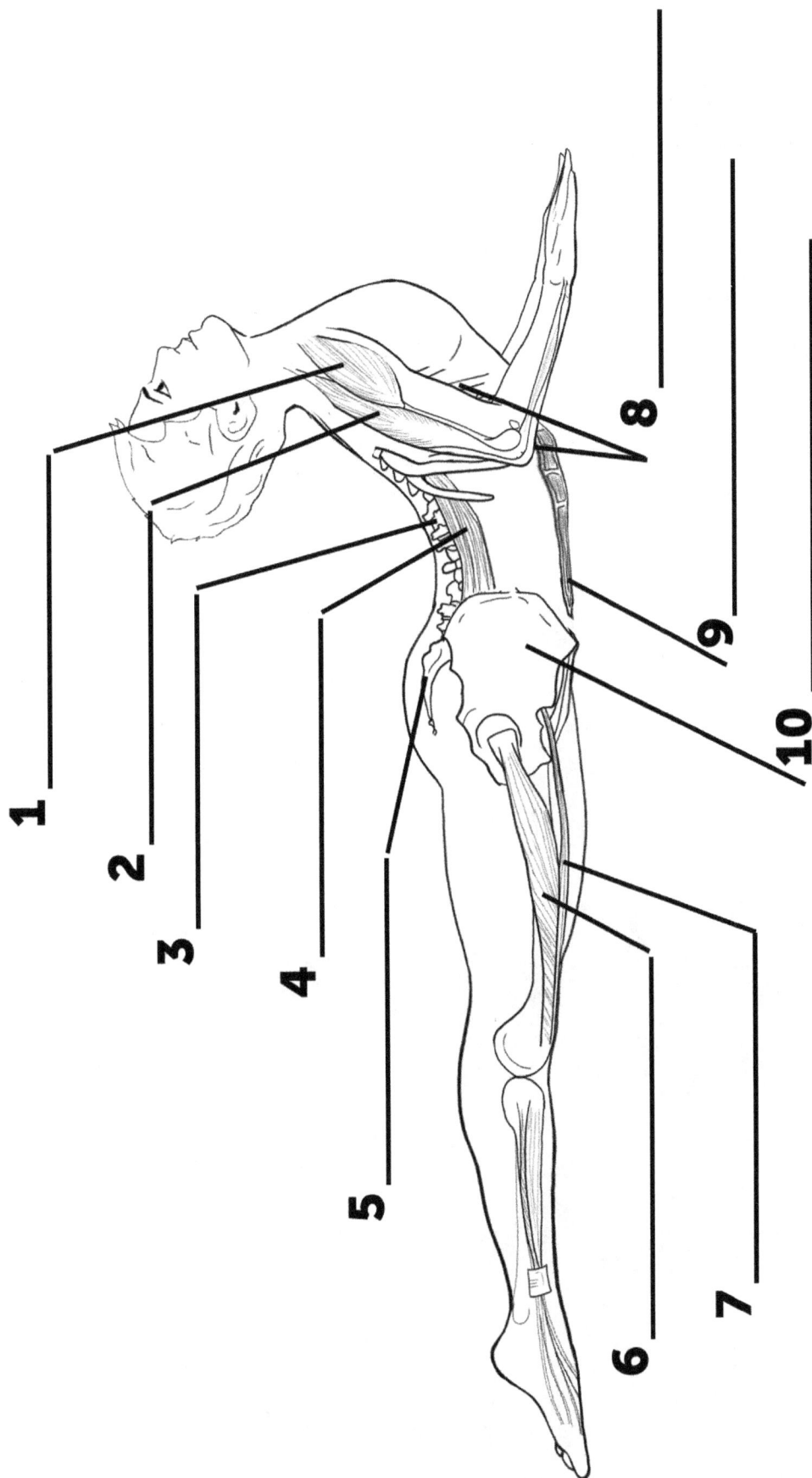

1

2

3

4

5

6

7

8

9

10

19. POSTURA DE LA COBRA

1.   DELTOIDES
2.   TRÍCEPS BRAQUIAL
3.   COLUMNA VERTEBRAL
4.   ERECTOR DE LA COLUMNA
5.   SACRO
6.   RECTO FEMORAL
7.   SARTORIO
8.   COSTILLAS
9.   RECTO ABDOMINAL
10. PELVIS

20. PADANGUSTHASANA

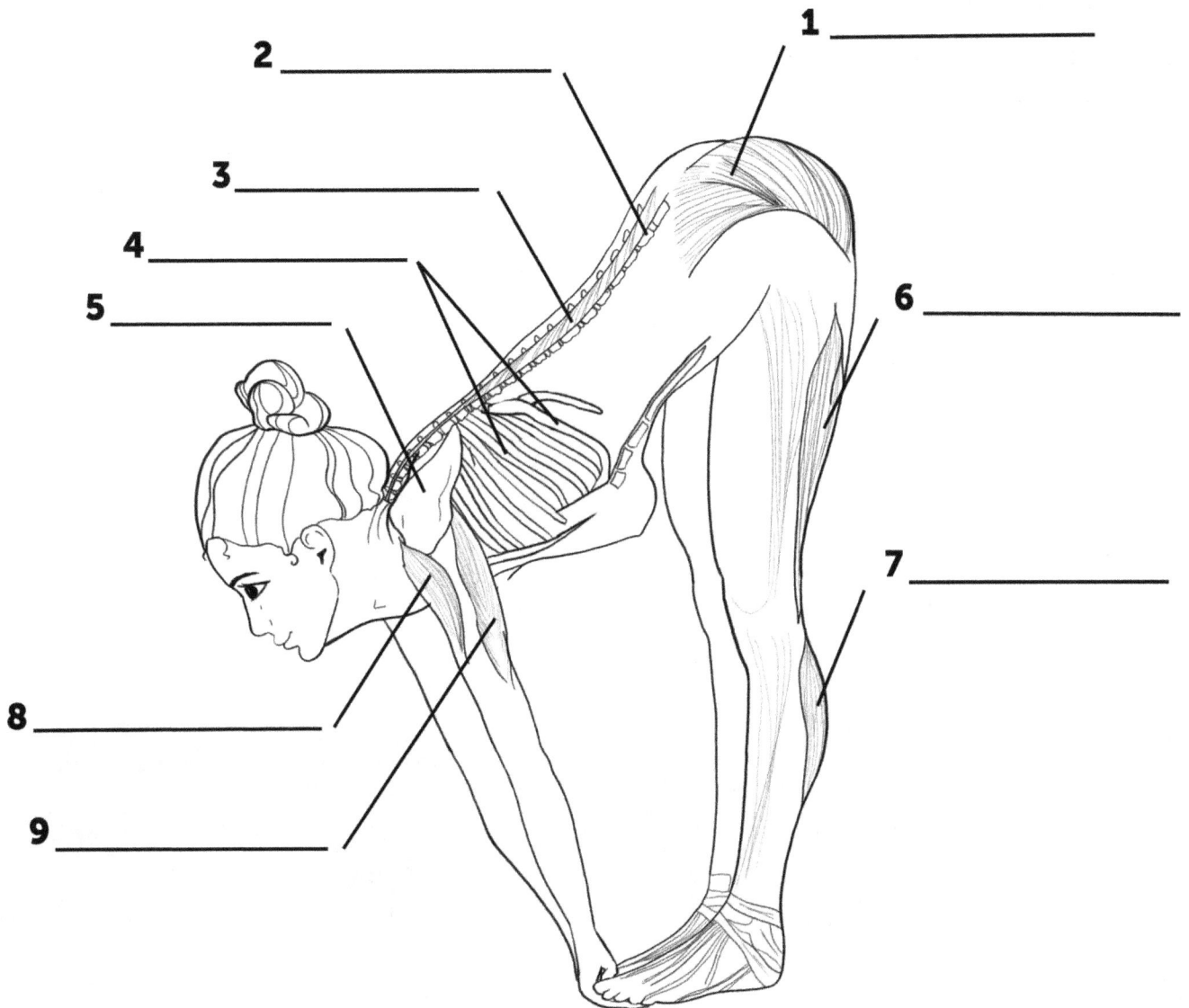

1 _____

2 _____

3 _____

4 _____

5 _____

6 _____

7 _____

8 _____

9 _____

20. PADANGUSTHASANA

1. PIRIFORME
2. COLUMNA VERTEBRAL
3. MÚSCULOS ESPINALES
4. COSTILLAS
5. ESCÁPULA
6. ISQUIOTIBIALES
7. GASTROCNEMIO
8. DELTOIDES
9. TRÍCEPS BRAQUIAL

21. POSTURA DEL NIÑO

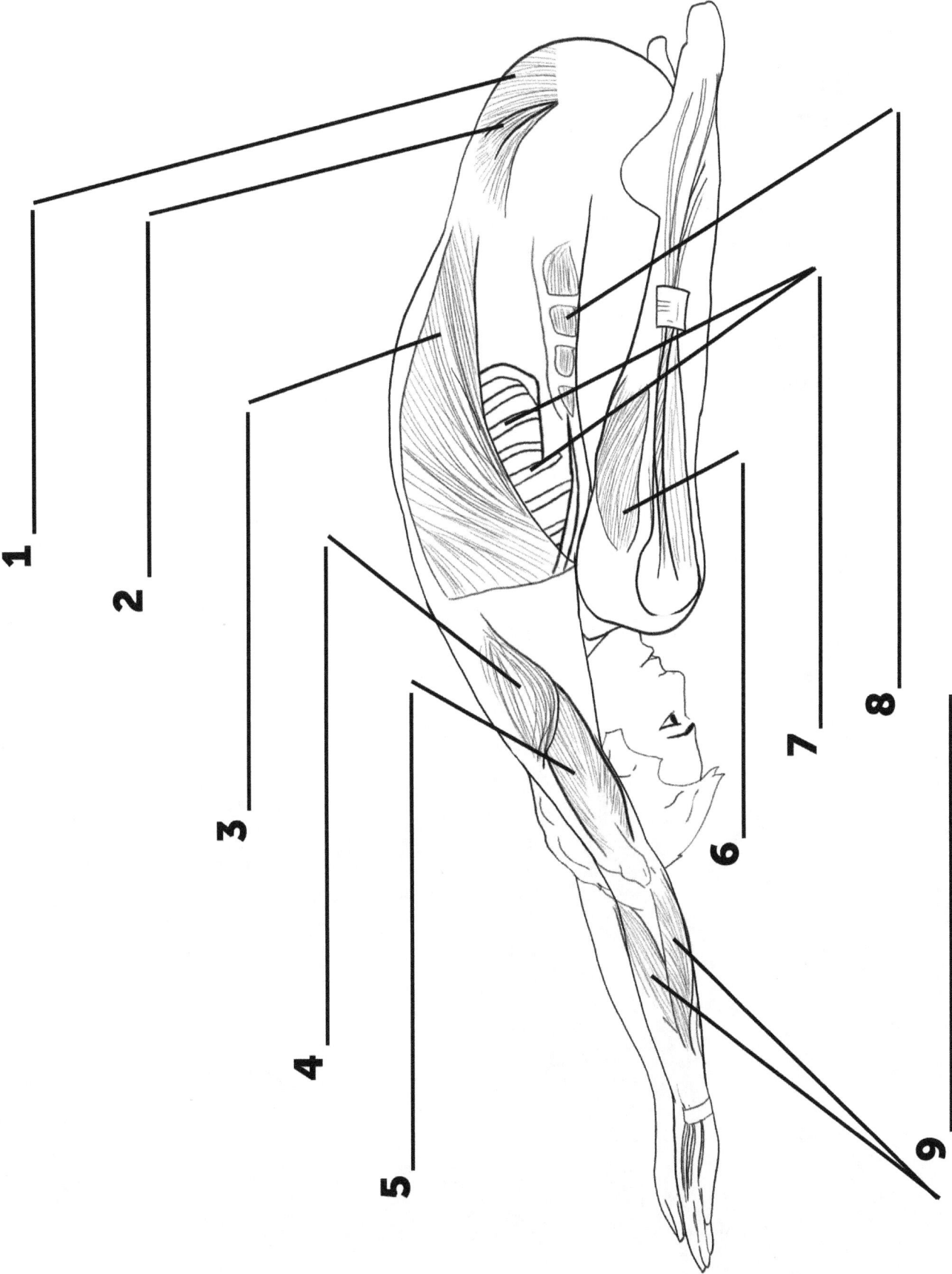

1

2

3

4

5

6

7

8

9

21. POSTURA DEL NIÑO

1. MÚSCULO GLÚTEO MAYOR

2. PIRIFORME

3. LATISSIMUS DORSI

4. DELTOIDES

5. TRÍCEPS BRAQUIAL

6. GASTROCNEMIO

7. COSTILLAS

8. RECTO ABDOMINAL

9. PRONADORES

22. POSTURA DO BARCO

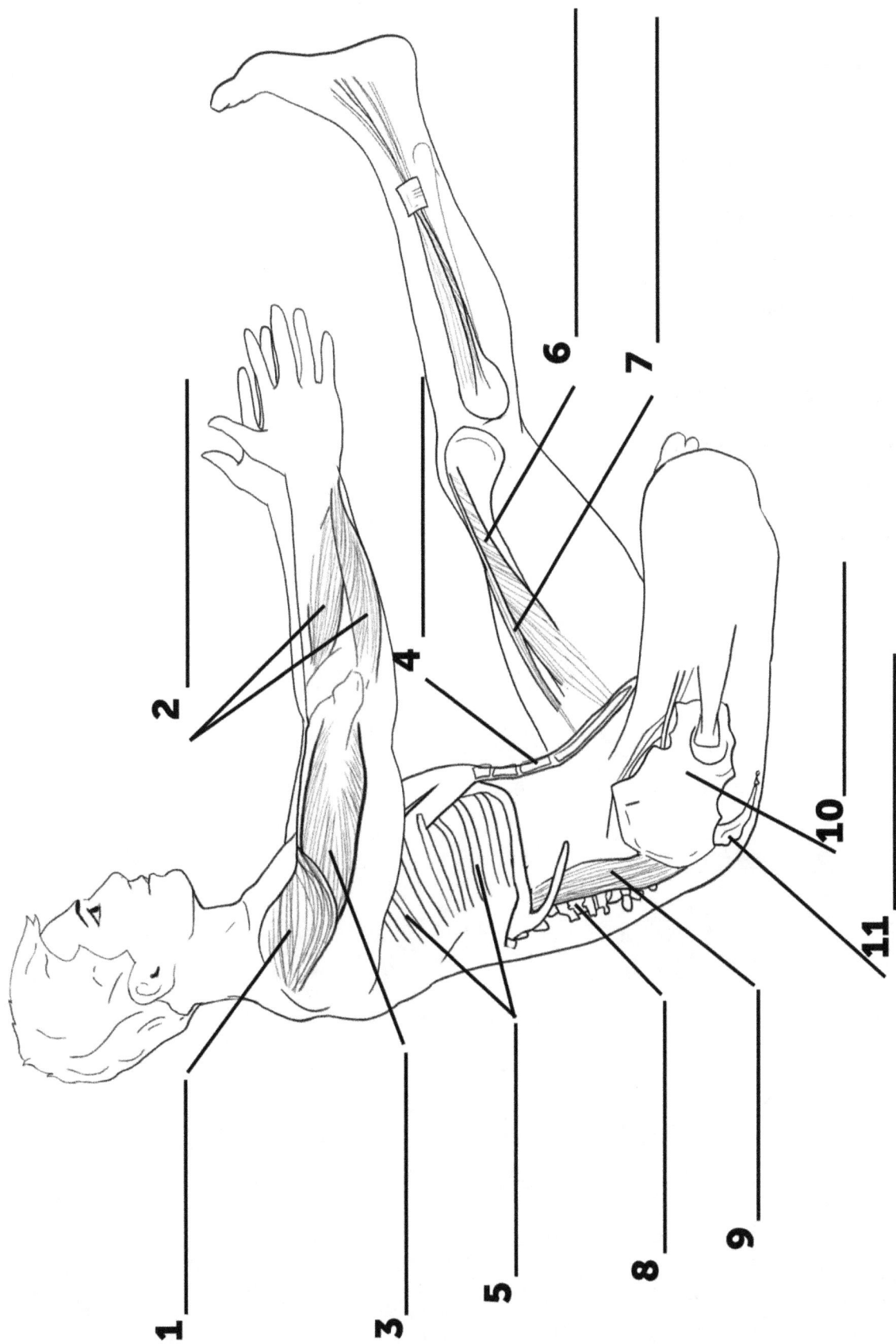

1

2

3

4

5

6

7

8

9

10

11

22. POSTURA DO BARCO

1. DELTOIDES

2. PRONADORES

3. TRÍCEPS BRAQUIAL

4. RECTO ABDOMINAL

5. COSTILLAS

6. RECTO FEMORAL

7. SARTORIO

8. COLUMNA VERTEBRAL

9. ERECTOR DE LA COLUMNA

10. PELVIS

11. SACRO

23. POSTURA DEL DELFÍN

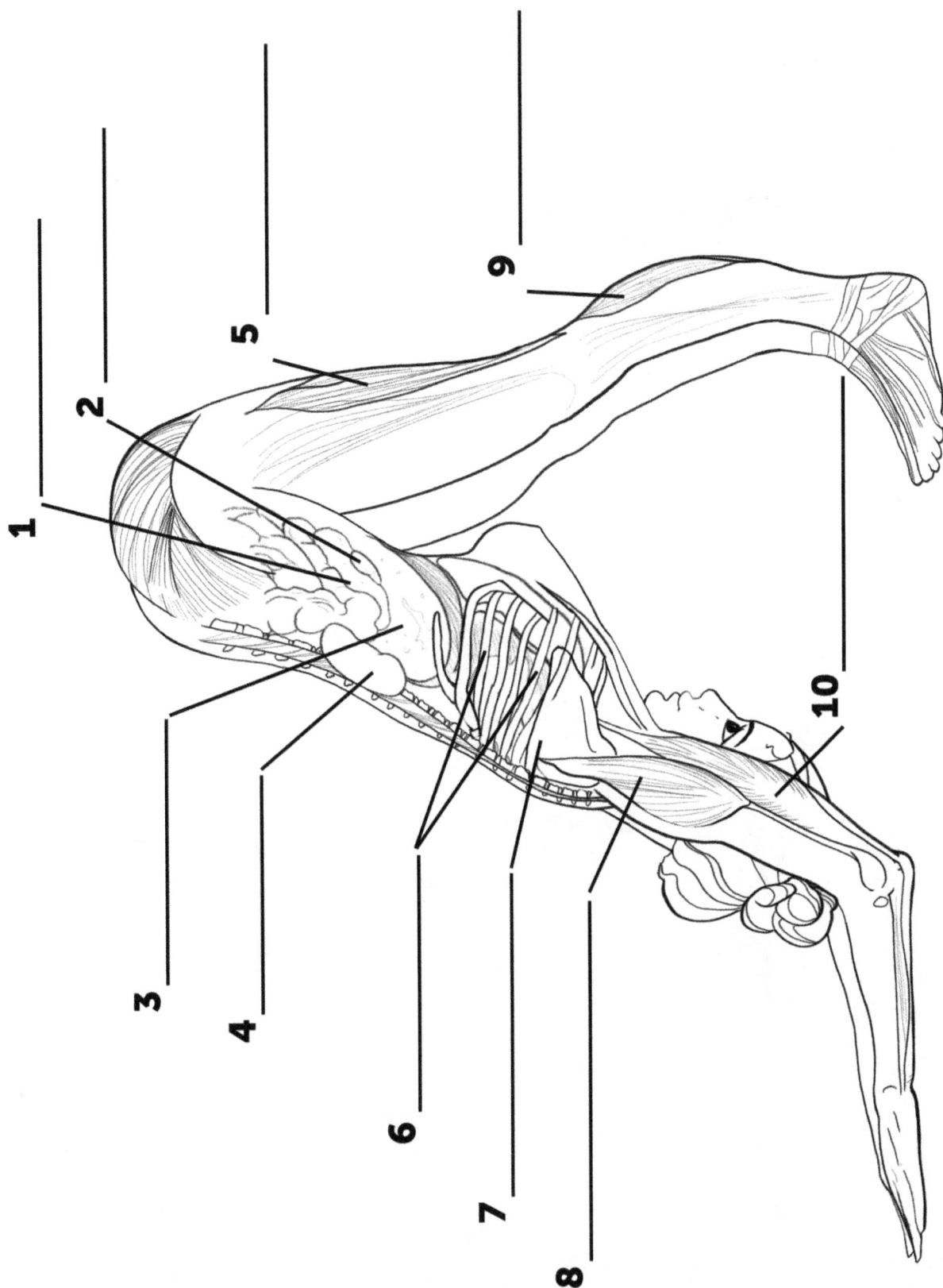

23. POSTURA DEL DELFÍN

1. ESTÓMAGO
2. VESÍCULA BILIAR
3. HÍGADO
4. RIÑÓN
5. ISQUIOTIBIALES
6. COSTILLAS
7. ESCÁPULA
8. DELTOIDES
9. GASTROCNEMIO
10. TRÍCEPS BRAQUIAL

24. POSTURA DEL PUENTE

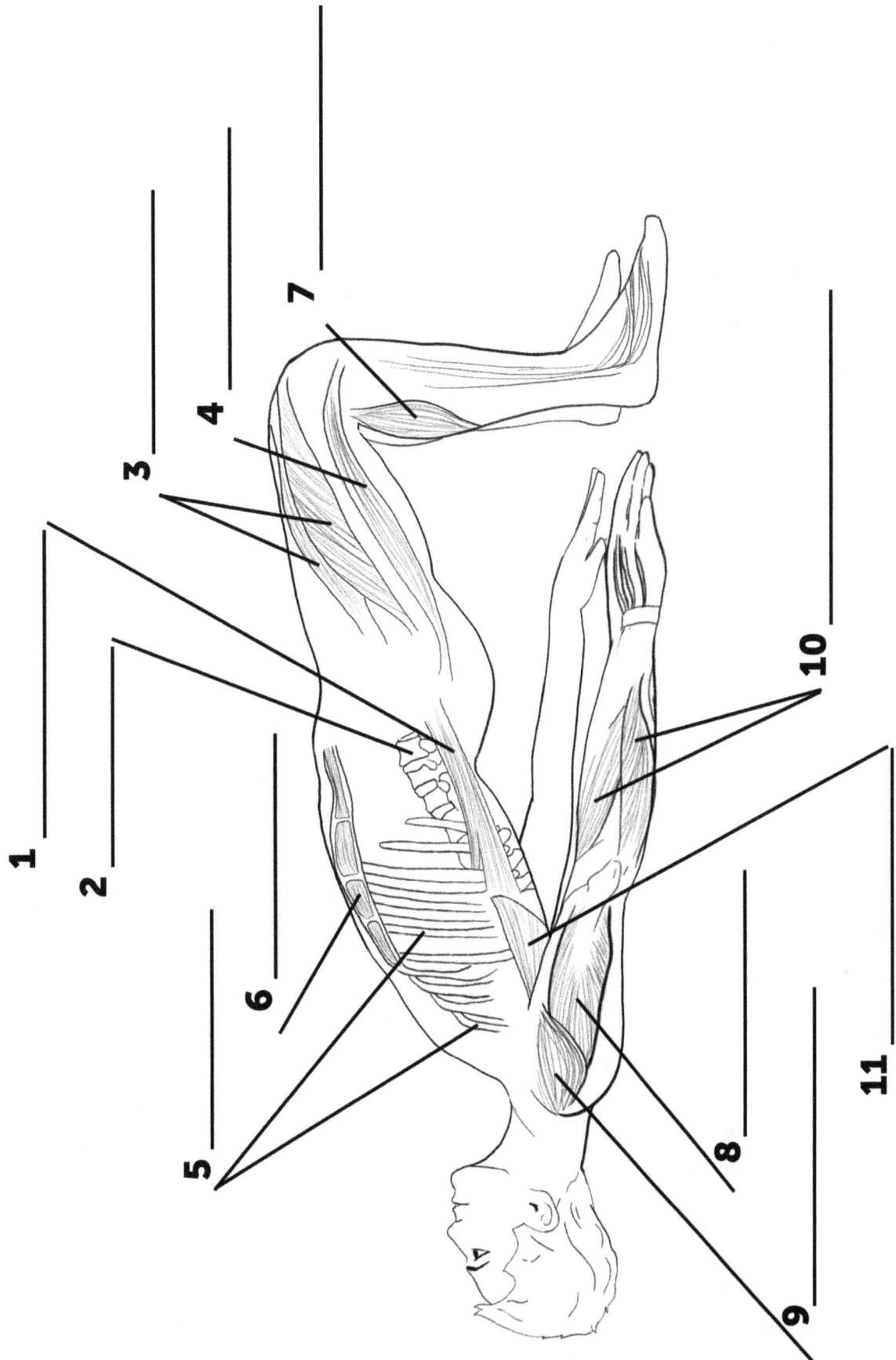

1

2

3

4

5

6

7

8

9

10

11

24. POSTURA DEL PUENTE

1. ERECTOR DE LA COLUMNA
2. COLUMNA VERTEBRAL
3. CUADRÍCEPS
4. ISQUIOTIBIALES
5. COSTILLAS
6. RECTO ABDOMINAL
7. GASTROCNEMIO
8. TRÍCEPS BRAQUIAL
9. DELTOIDES
10. PRONADORES
11. INFRAESPINOSO

25. POSTURA DE LA GUIRNALDA

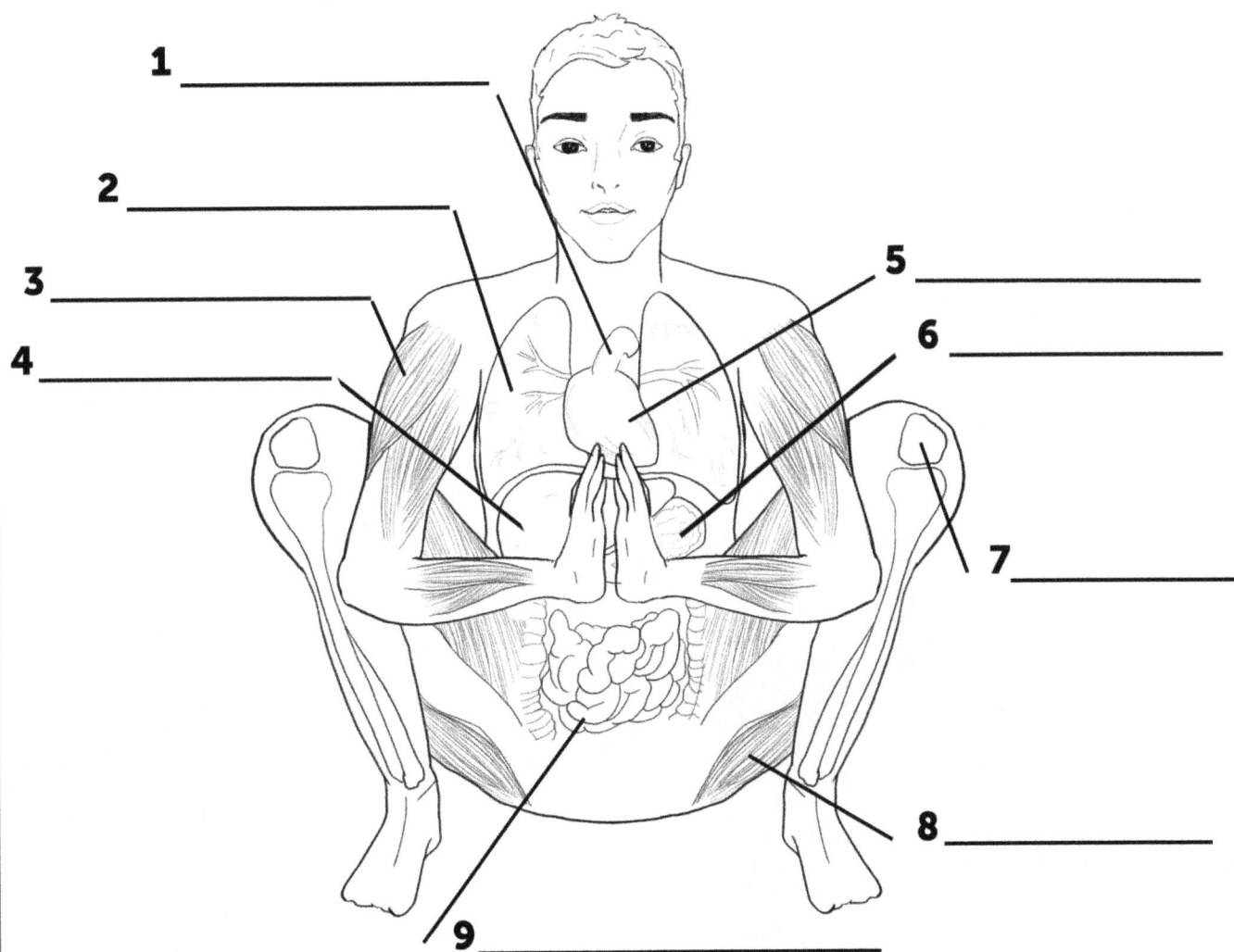

1 _____

2 _____

3 _____

4 _____

5 _____

6 _____

7 _____

8 _____

9 _____

25. POSTURA DE LA GUIRNALDA

1. AORTA
2. PULMONES
3. TRÍCEPS BRAQUIAL
4. HÍGADO
5. CORAZÓN
6. ESTÓMAGO
7. RÓTULA
8. ISQUIOTIBIALES
9. FOLICULOS DE INTESTINO DELGADO

26. POSTURA DEL PERRO

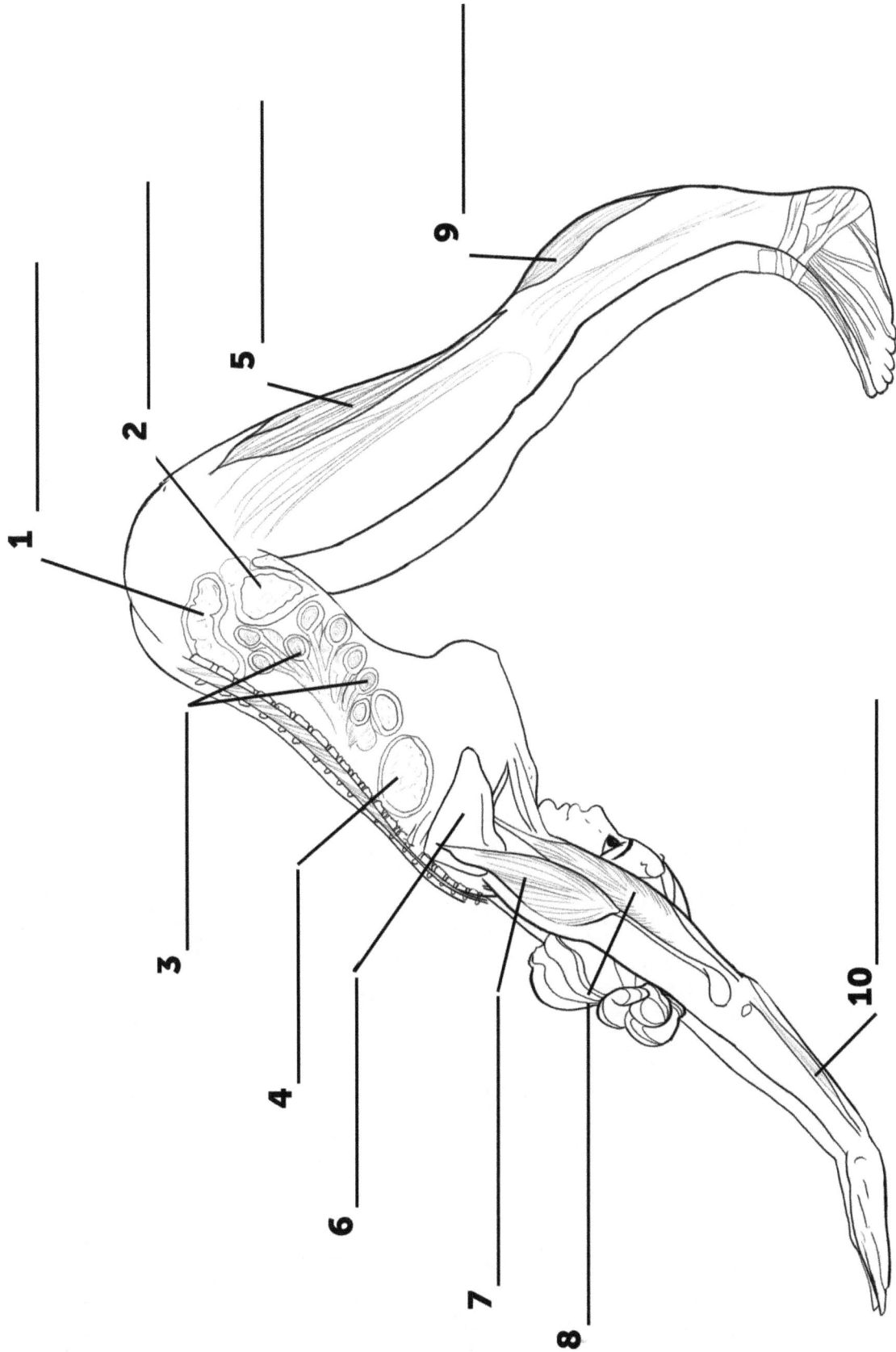

26. POSTURA DEL PERRO

1. RECTO
2. VEJIGA URINARIA
3. INTESTINO DELGADO
4. ESTÓMAGO
5. ISQUIOTIBIALES
6. ESCÁPULA
7. DELTOIDES
8. TRÍCEPS BRAQUIAL
9. GASTROCNEMIO
10. PRONADORES

27. POSTURA DEL TABLÓN

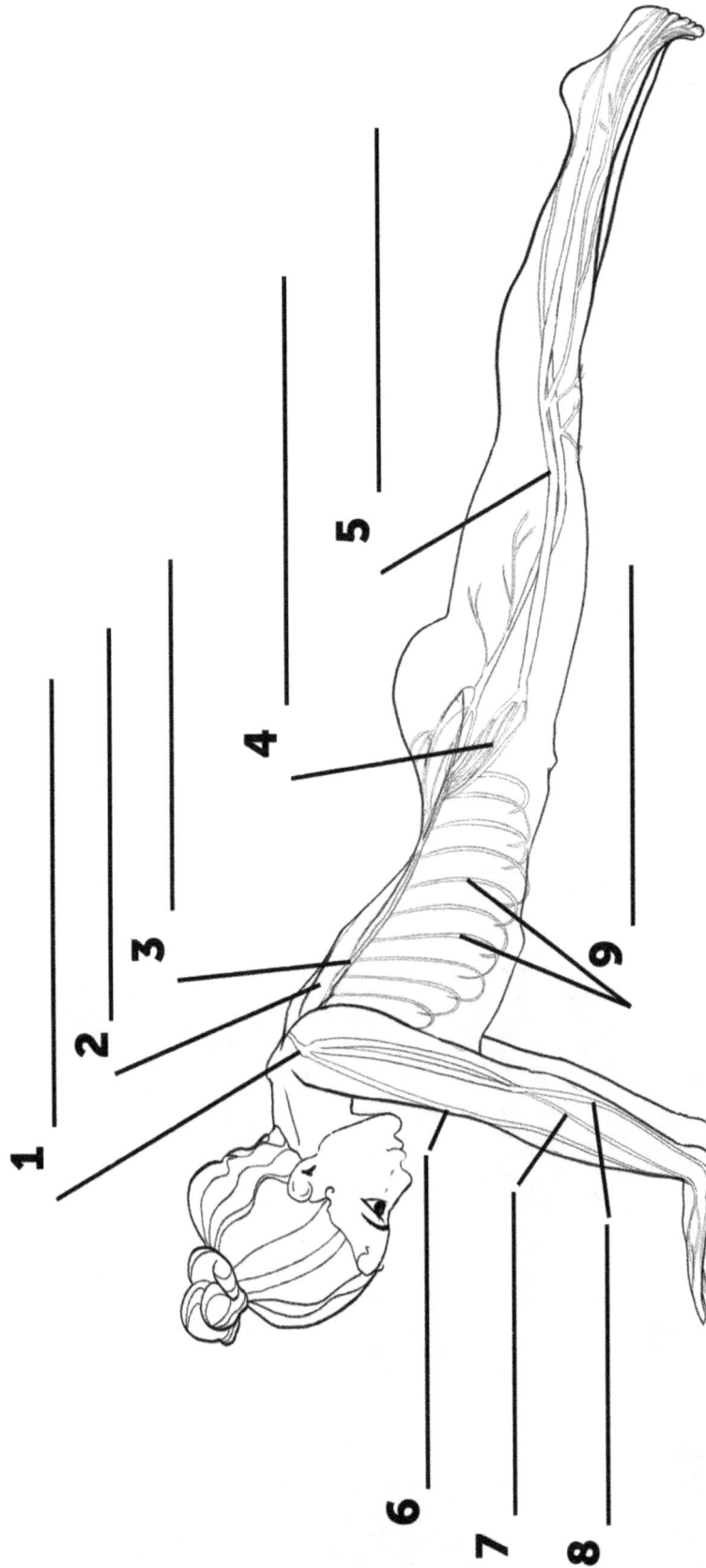

27. POSTURA DEL TABLÓN

1. PLEXO BRAQUIAL
2. MÉDULA ESPINAL
3. NERVIO VAGO
4. PLEXO LUMBAR
5. CIÁTICO
6. ULNAR
7. MEDIANA
8. RADIAL
9. INTERCOSTALES

28. CHATURANGA

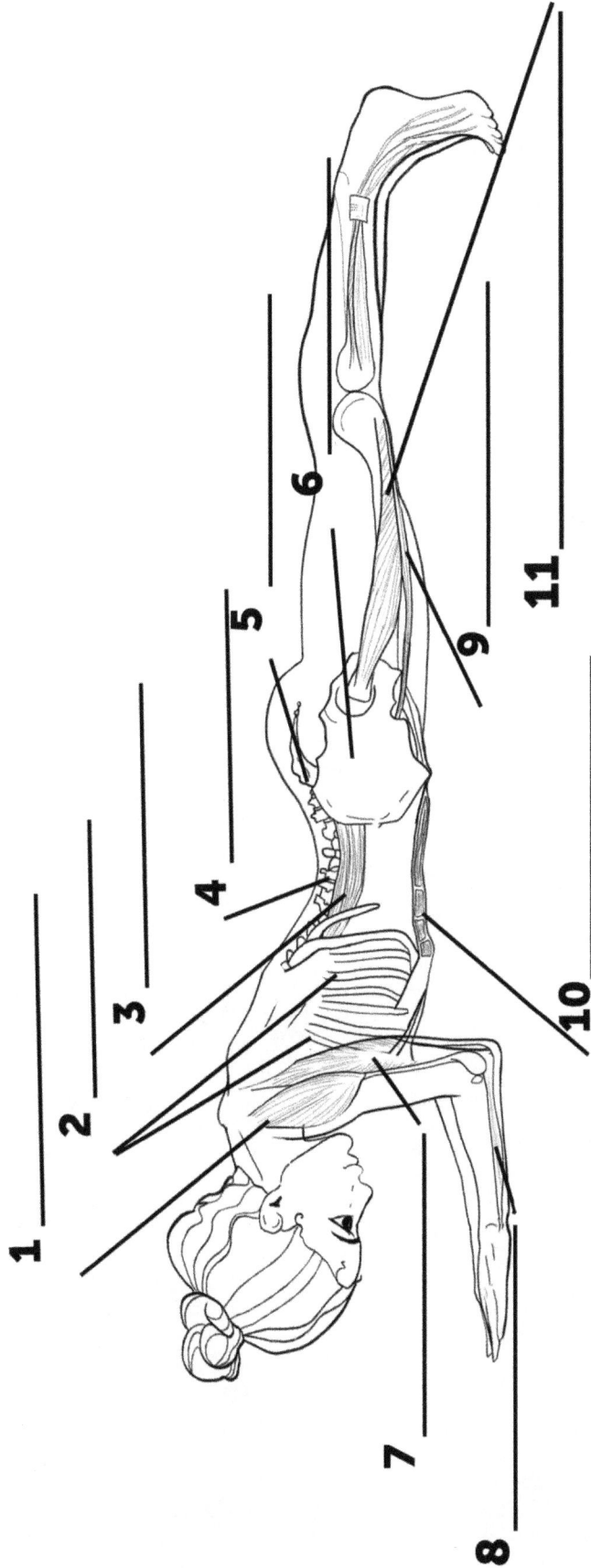

1

2

3

4

5

6

7

8

9

10

11

28. CHATURANGA

1. DELTOIDES
2. COSTILLAS
3. ERECTOR DE LA COLUMNA
4. COLUMNA VERTEBRAL
5. SACRO
6. PELVIS
7. TRÍCEPS BRAQUIAL
8. PRONADORES
9. SARTORIO
10. RECTO ABDOMINAL
11. RECTO FEMORAL

29. KAPOTASANA

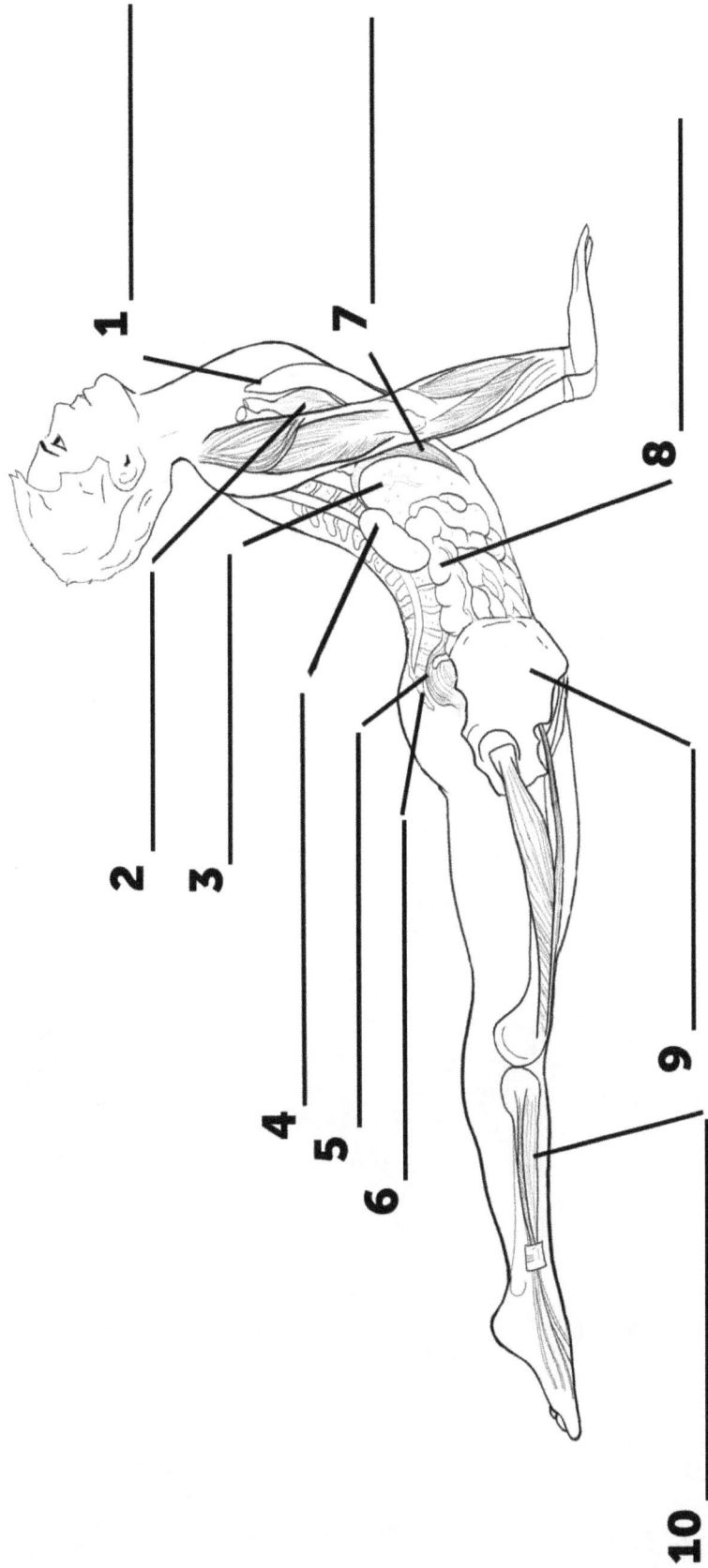

1

7

2

3

4

5

6

8

9

10

29. KAPOTASANA

1.   PULMONES
2.   CORAZÓN
3.   HÍGADO
4.   RIÑÓN
5.   RECTO
6.   SACRO
7.   DIAFRAGMA
8.   COLON ASCENDENTE
9.   PELVIS
10.   TIBIAL ANTERIOR

30. PAVANAMUKTASANA.

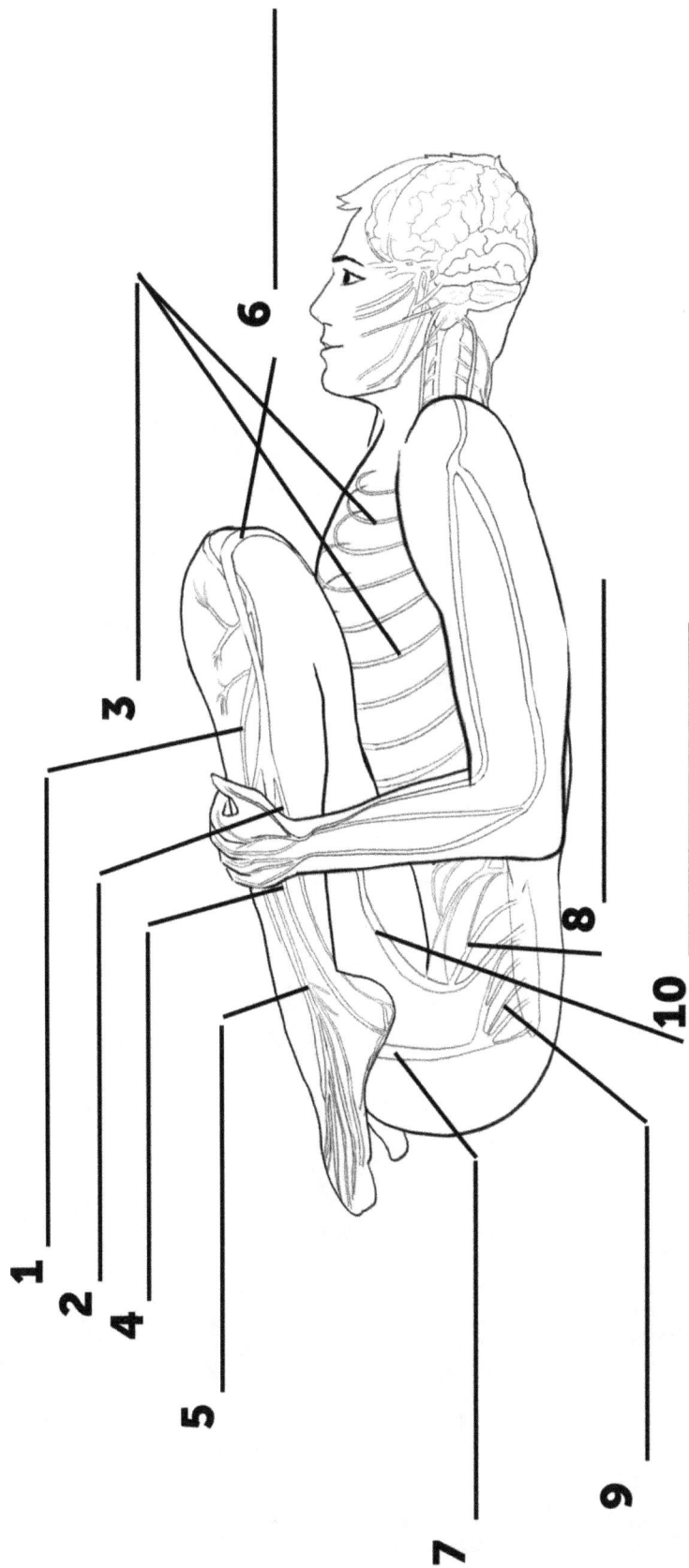

1

2

3

4

5

6

7

8

9

10

30. PAVANAMUKTASANA.

1. SAFENA
2. PERONEO COMÚN
3. INTERCOSTALES
4. TIBIAL
5. PERONEO SUPERFICIAL
6. CIÁTICA
7. CIÁTICA
8. PLEXO LUMBAR
9. PLEXO SACRO
10. FEMORAL

31. UTTANPADASANA

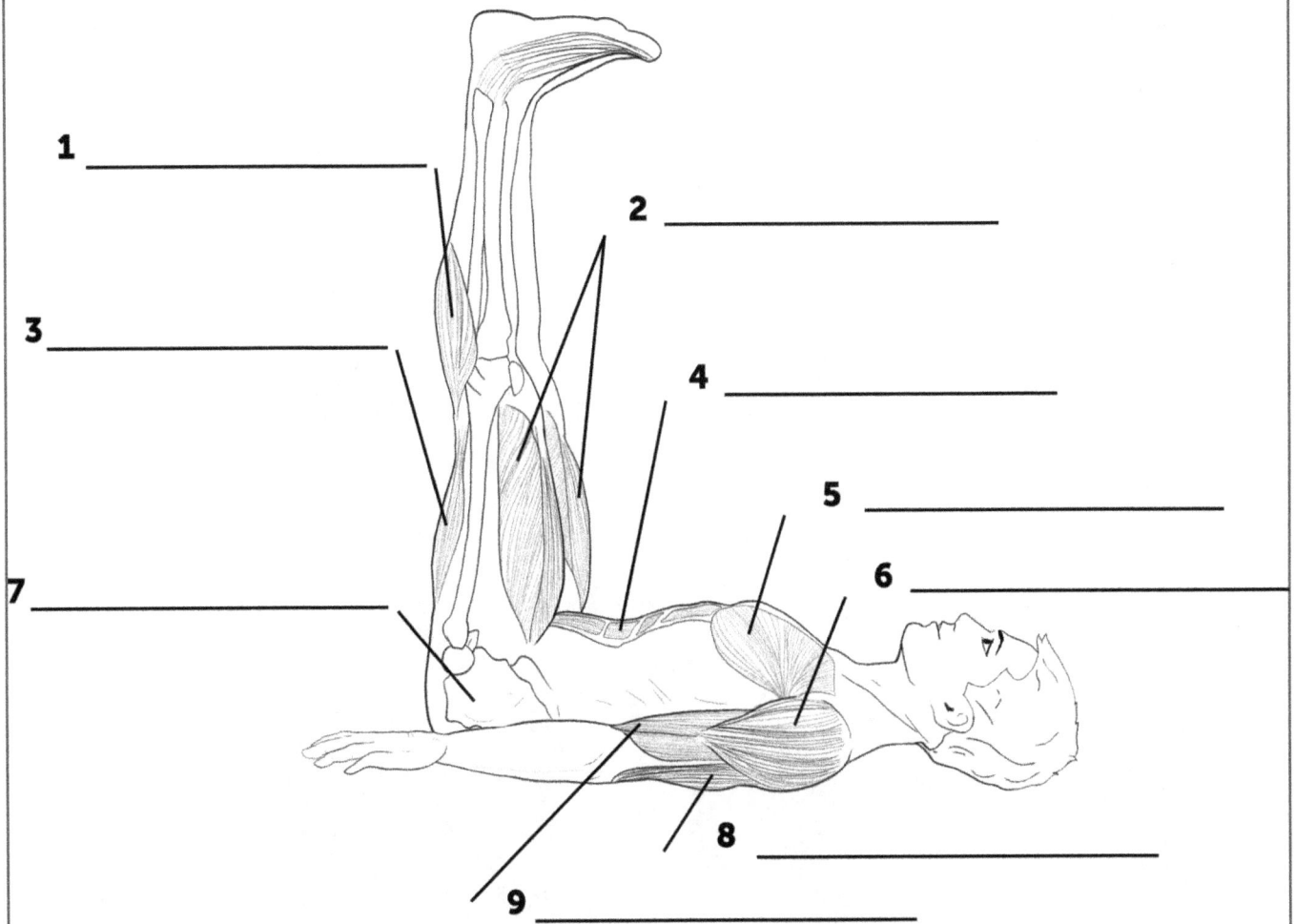

1 _____

2 _____

3 _____

4 _____

5 _____

6 _____

7 _____

8 _____

9 _____

31. UTTANPADASANA

1. GASTROCNEMIO
2. CUADRÍCEPS
3. ISQUIOTIBIALES
4. RECTO ABDOMINAL
5. PECTORAL MAYOR
6. DELTOIDES
7. PELVIS
8. TRÍCEPS BRAQUIAL
9. BÍCEPS BRAQUIAL

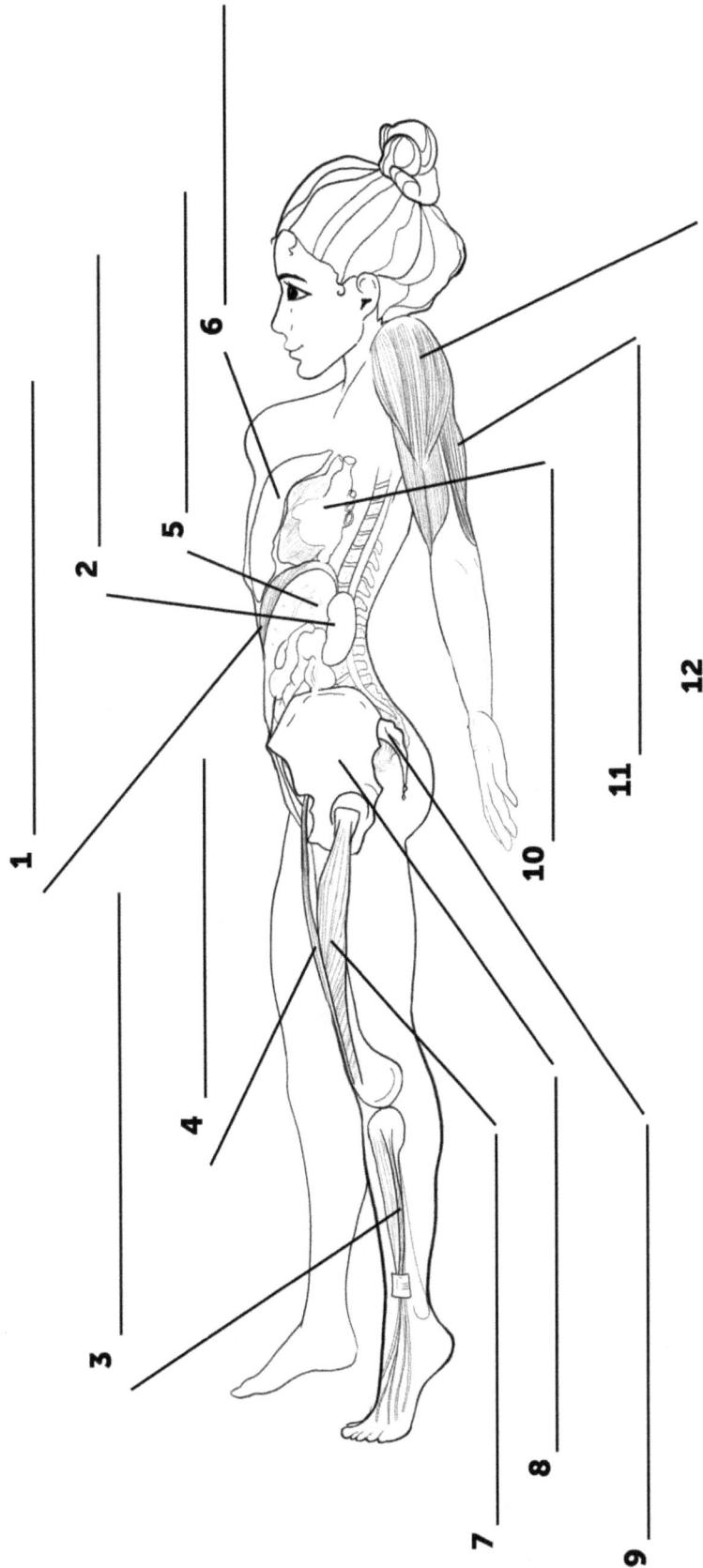

32. SHAVASANA

1

2

3

4

5

6

7

8

9

10

11

12

32. SHAVASANA

1.   DIAFRAGMA

2.   RIÑÓN

3.   TIBIAL ANTERIOR

4.   SARTORIO

5.   HÍGADO

6.   PULMONES

7.   RECTO FEMORAL

8.   PELVIS

9.   SACRO

10. CORAZÓN

11. TRÍCEPS BRAQUIAL

12. DELTOIDES

33. HASTA UTTANASANA

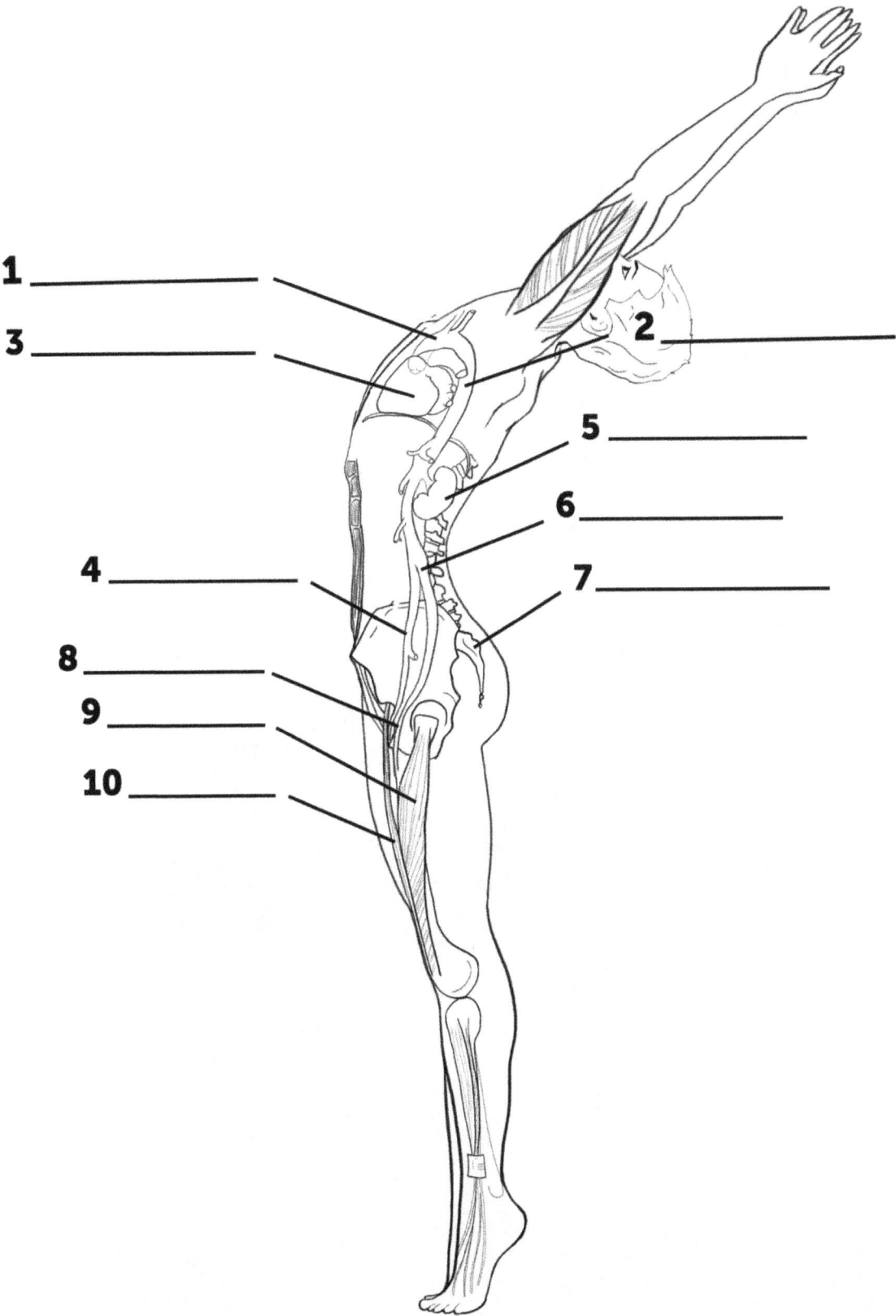

1 _____

3 _____

2 _____

5 _____

6 _____

4 _____

7 _____

8 _____

9 _____

10 _____

33. HASTA UTTANASANA

1. AORTA TORÁCICA ASCENDENTE
2. AORTA TORÁCICA DESCENDENTE
3. CORAZÓN
4. ARTERIA ILIACA COMÚN
5. RIÑÓN
6. AORTA ABDOMINAL
7. SACRO
8. ARTERIA FEMORAL
9. RECTO FEMORAL
10. SARTORIO

34. POSTURA DE LA RANA

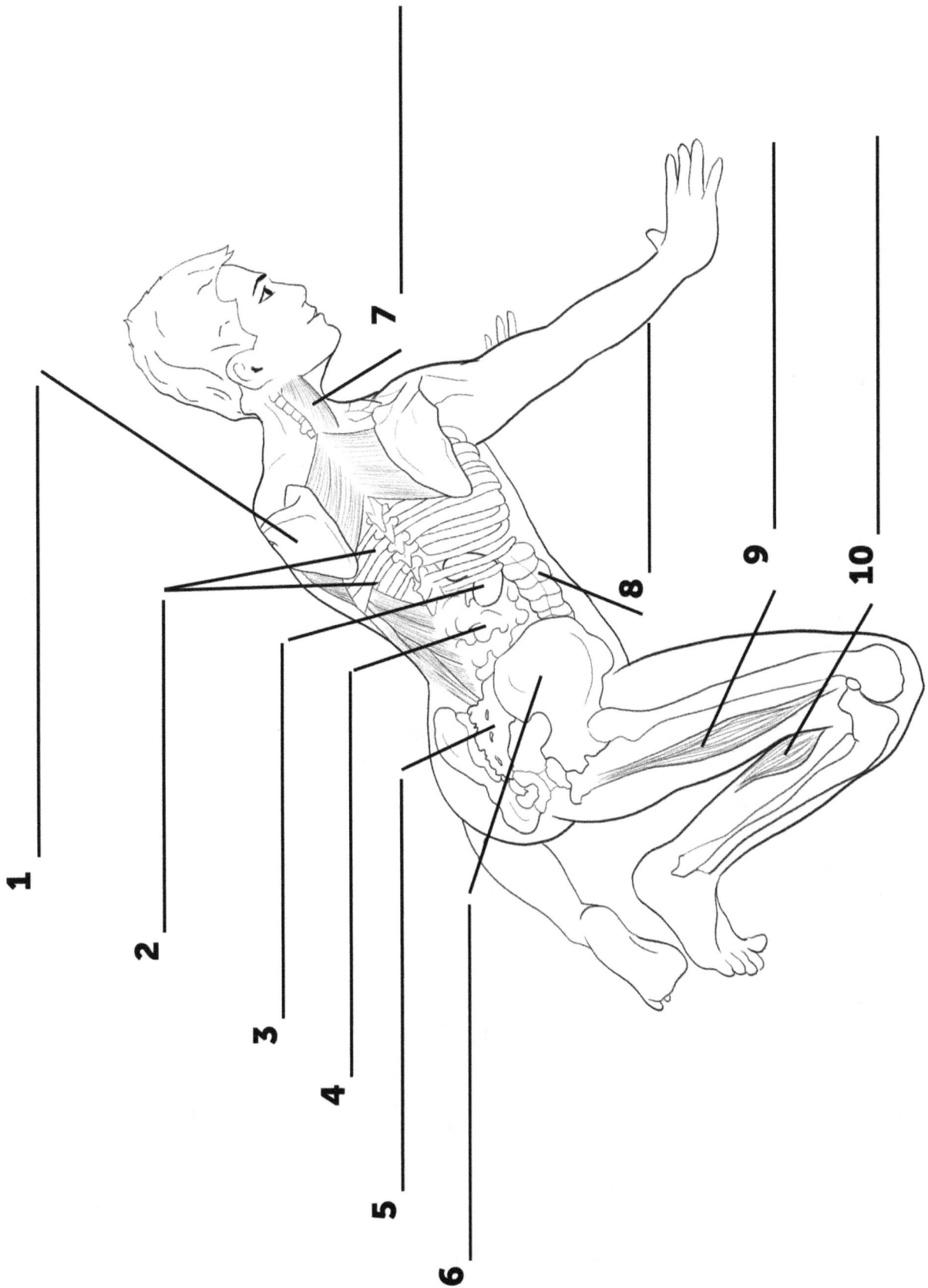

1

2

3

4

5

6

7

8

9

10

34. POSTURA DE LA RANA

1. ESCÁPULA

2. COSTILLAS

3. RIÑÓN

4. FOLICULOS DE INTESTINO DELGADO

5. SACRO

6. PELVIS

7. MÚSCULO ESPLENIO DE LA CABEZA

8. COLON ASCENDENTE

9. ISQUIOTIBIALES

10. GASTROCNEMIO

35. POSTURA DE MEDIO LOTO

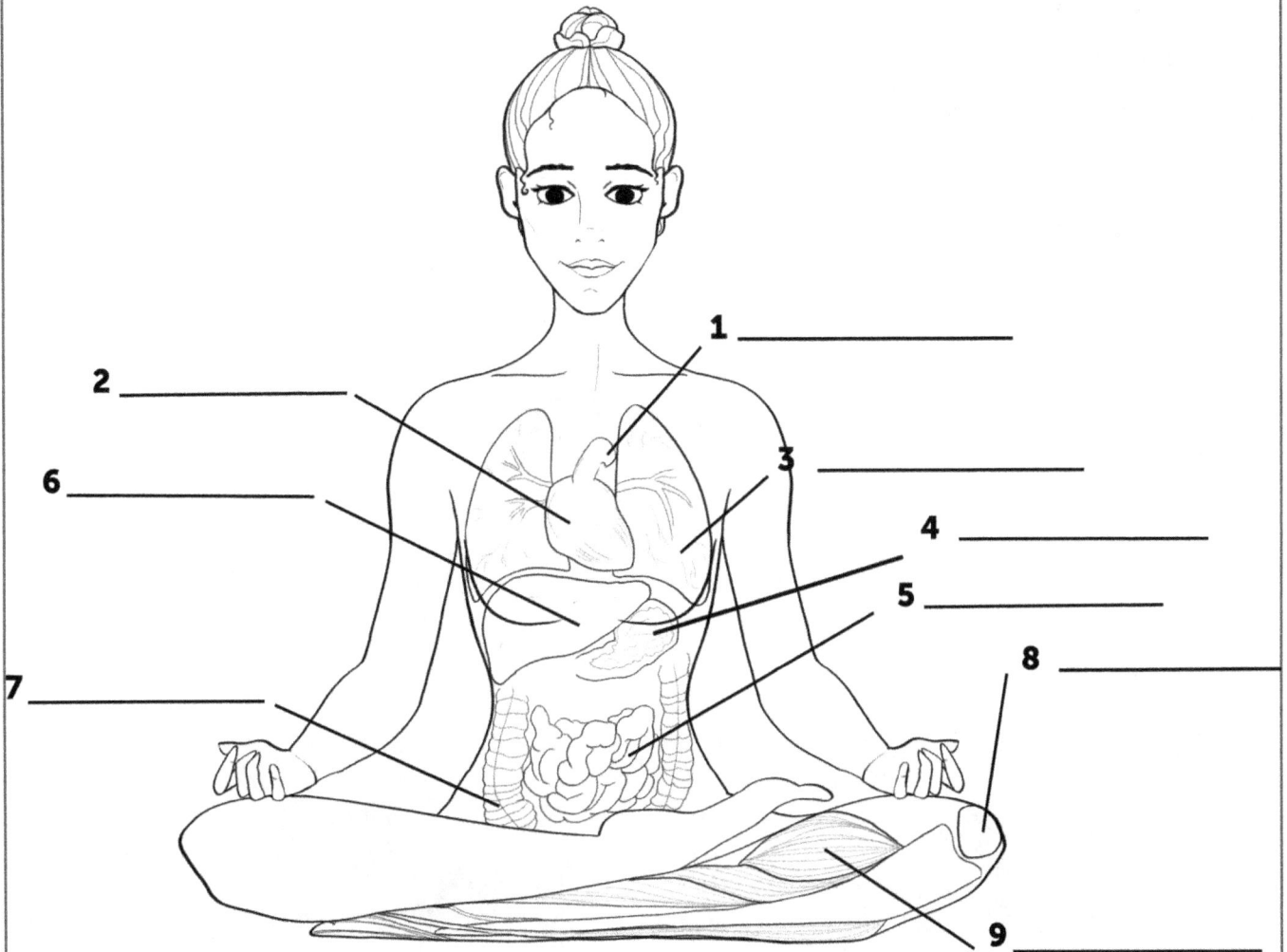

1 _____

2 _____

3 _____

4 _____

5 _____

6 _____

7 _____

8 _____

9 _____

35. POSTURA DE MEDIO LOTO

1.   AORTA

2.   CORAZÓN

3.   PULMONES

4.   ESTÓMAGO

5.   INTESTINO DELGADO

6.   HÍGADO

7.   INTESTINO GRUESO

8.   RÓTULA

9.   GASTROCNEMIO

36. BEBÉ FELIZ

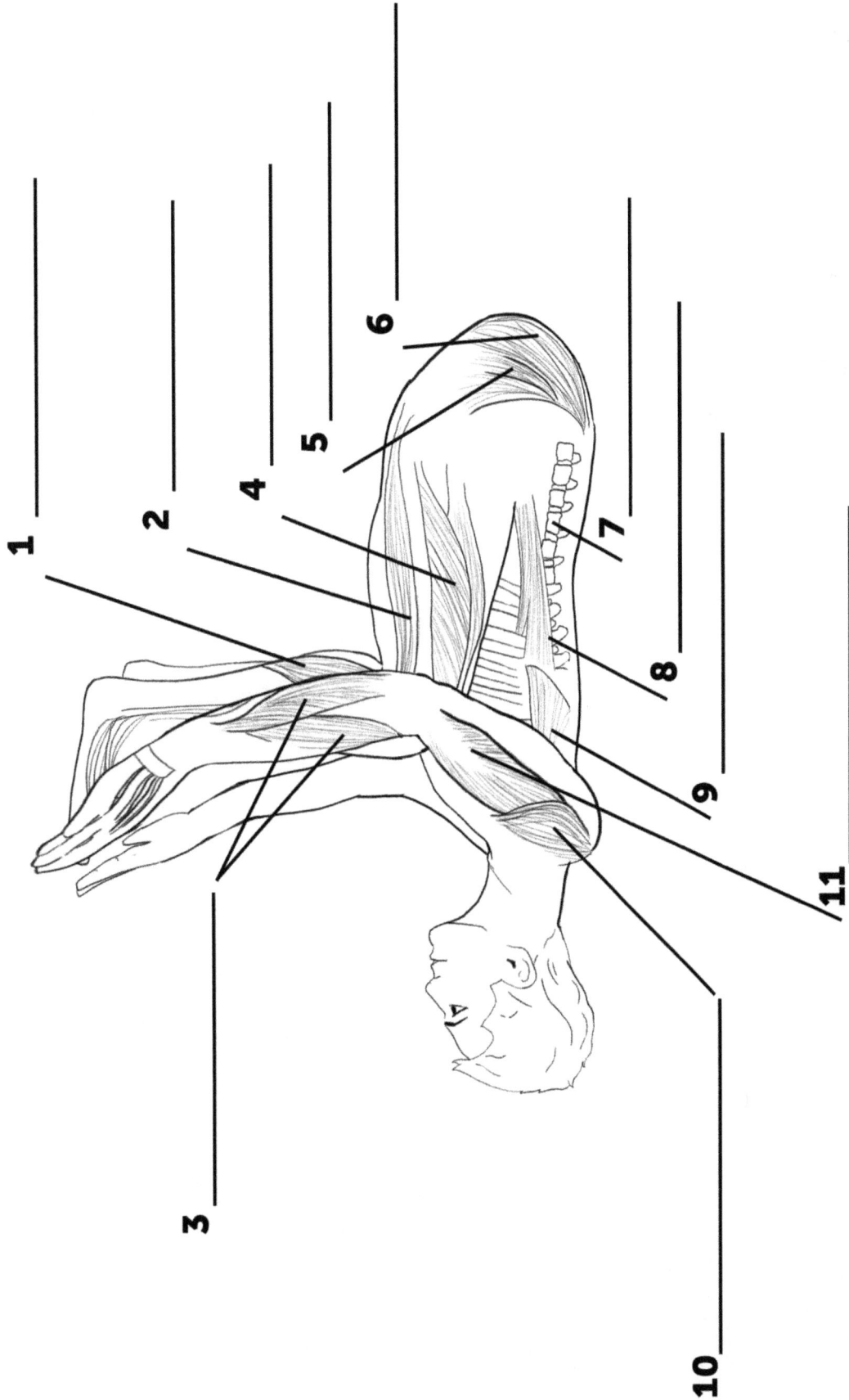

1
2
3
4
5
6
7
8
9
10
11

36. BEBÉ FELIZ

1. GASTROCNEMIO
2. ISQUIOTIBIALES
3. PRONADORES
4. CUADRÍCEPS
5. PIRIFORME
6. MÚSCULO GLÚTEO MAYOR
7. COLUMNA VERTEBRAL
8. ERECTOR DE LA COLUMNA
9. INFRAESPINOSO
10. DELTOIDES
11. TRÍCEPS BRAQUIAL

37. PIERNAS ARRIBA DE LA PARED

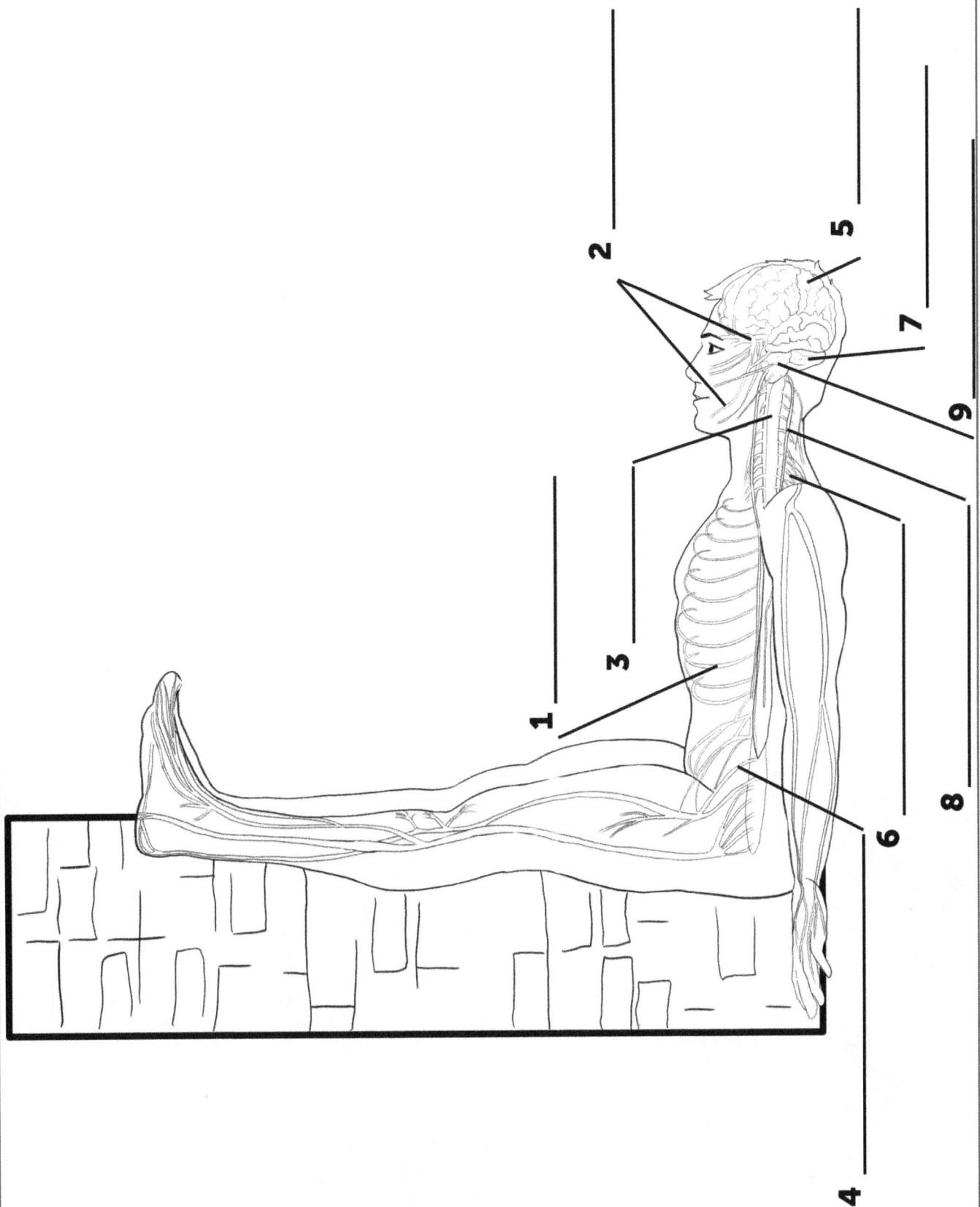

37. PIERNAS ARRIBA DE LA PARED

1. INTERCOSTALES
2. NERVIOS CRANEALES
3. MÉDULA ESPINAL
4. PLEXO LUMBAR
5. CEREBRO
6. PLEXO BRAQUIAL
7. CEREBELO
8. NERVIO VAGO
9. TRONCO ENCEFÁLICO

38. POSTURA KAPALBHATI

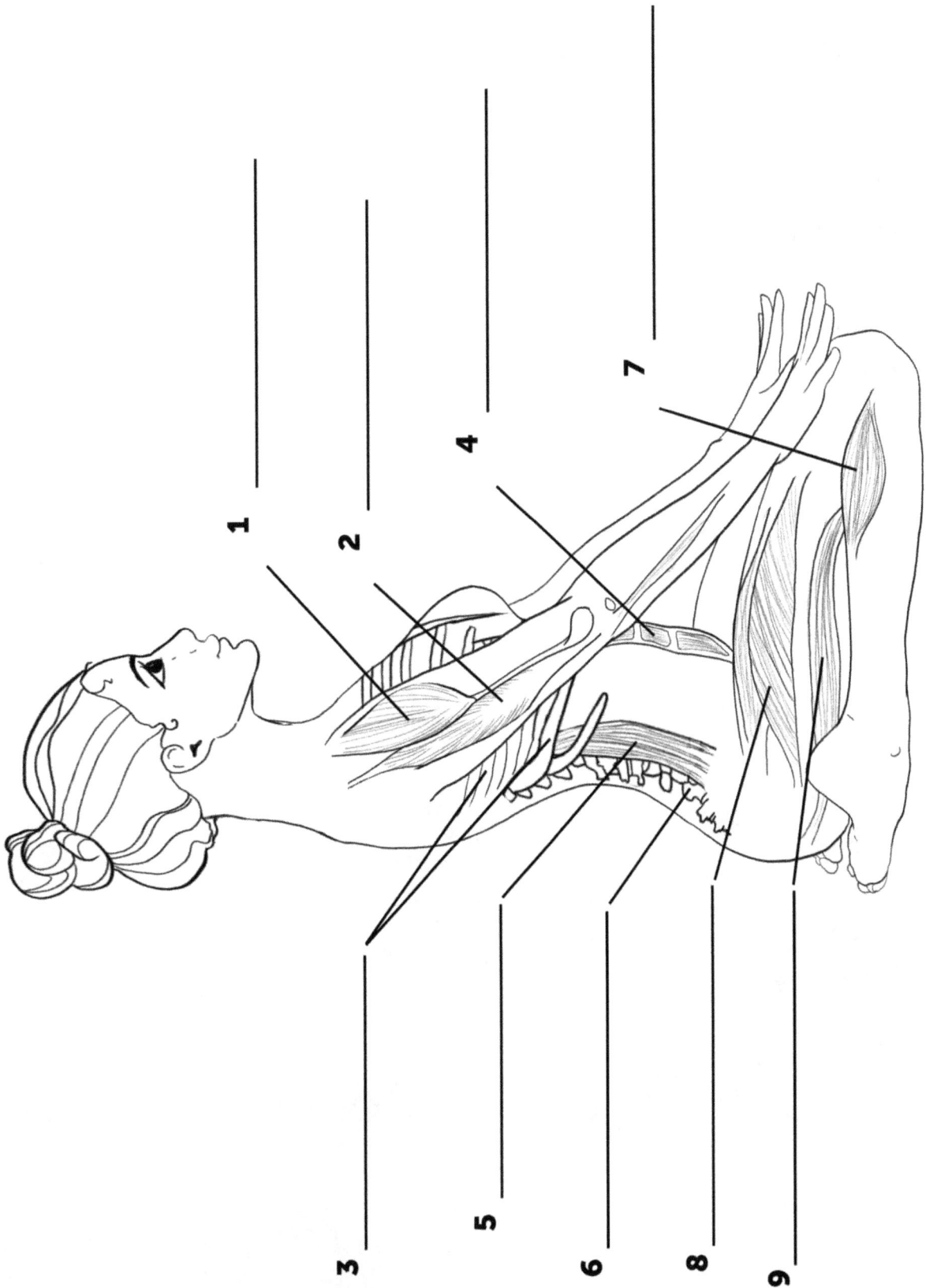

1

2

3

4

5

6

7

8

9

38. POSTURA KAPALBHATI

1. DELTOIDES
2. TRÍCEPS BRAQUIAL
3. COSTILLAS
4. RECTO ABDOMINAL
5. ERECTOR DE LA COLUMNA
6. COLUMNA VERTEBRAL
7. GASTROCNEMIO
8. CUADRÍCEPS
9. ISQUIOTIBIALES

39. POSE LOCUST

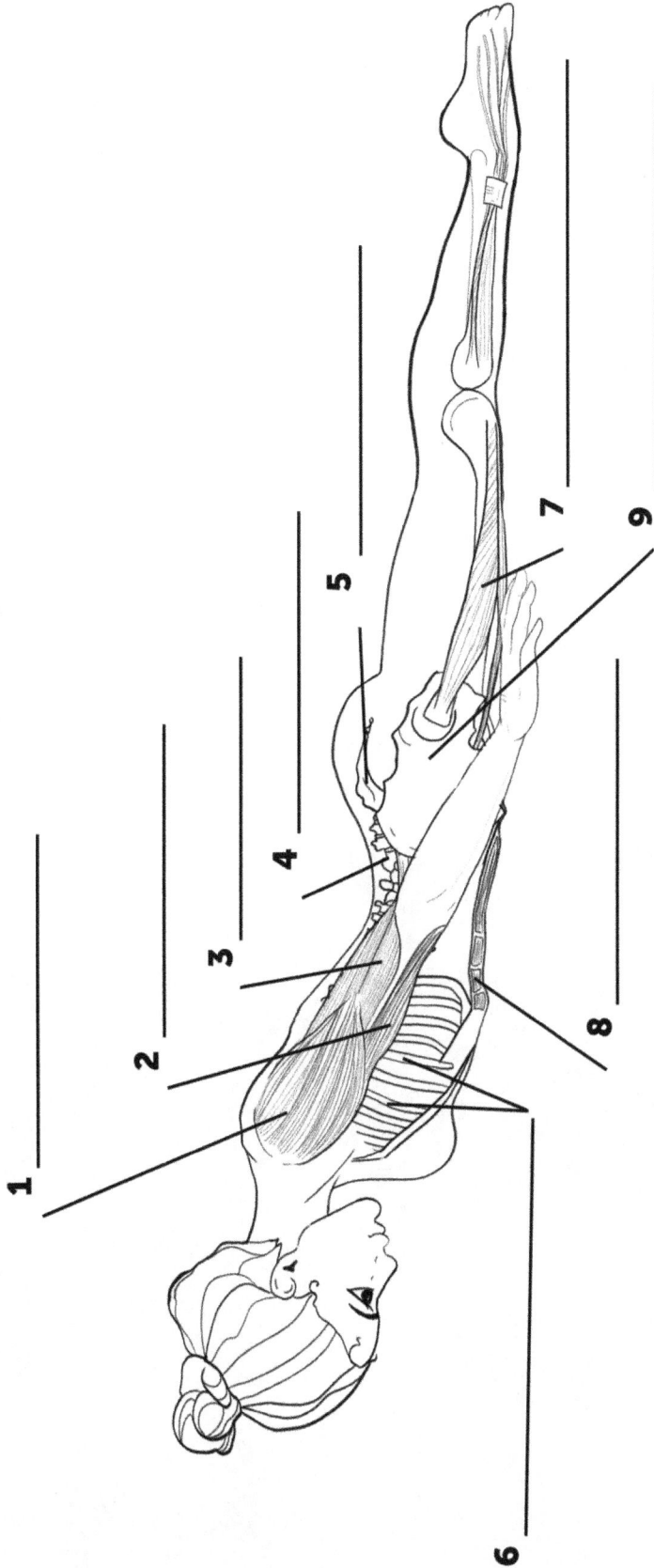

39. POSE LOCUST

1. DELTOIDES
2. BÍCEPS BRAQUIAL
3. TRÍCEPS BRAQUIAL
4. COLUMNA VERTEBRAL
5. SACRO
6. COSTILLAS
7. RECTO FEMORAL
8. RECTO ABDOMINAL
9. PELVIS

40. UTTANA SHISHOSANA

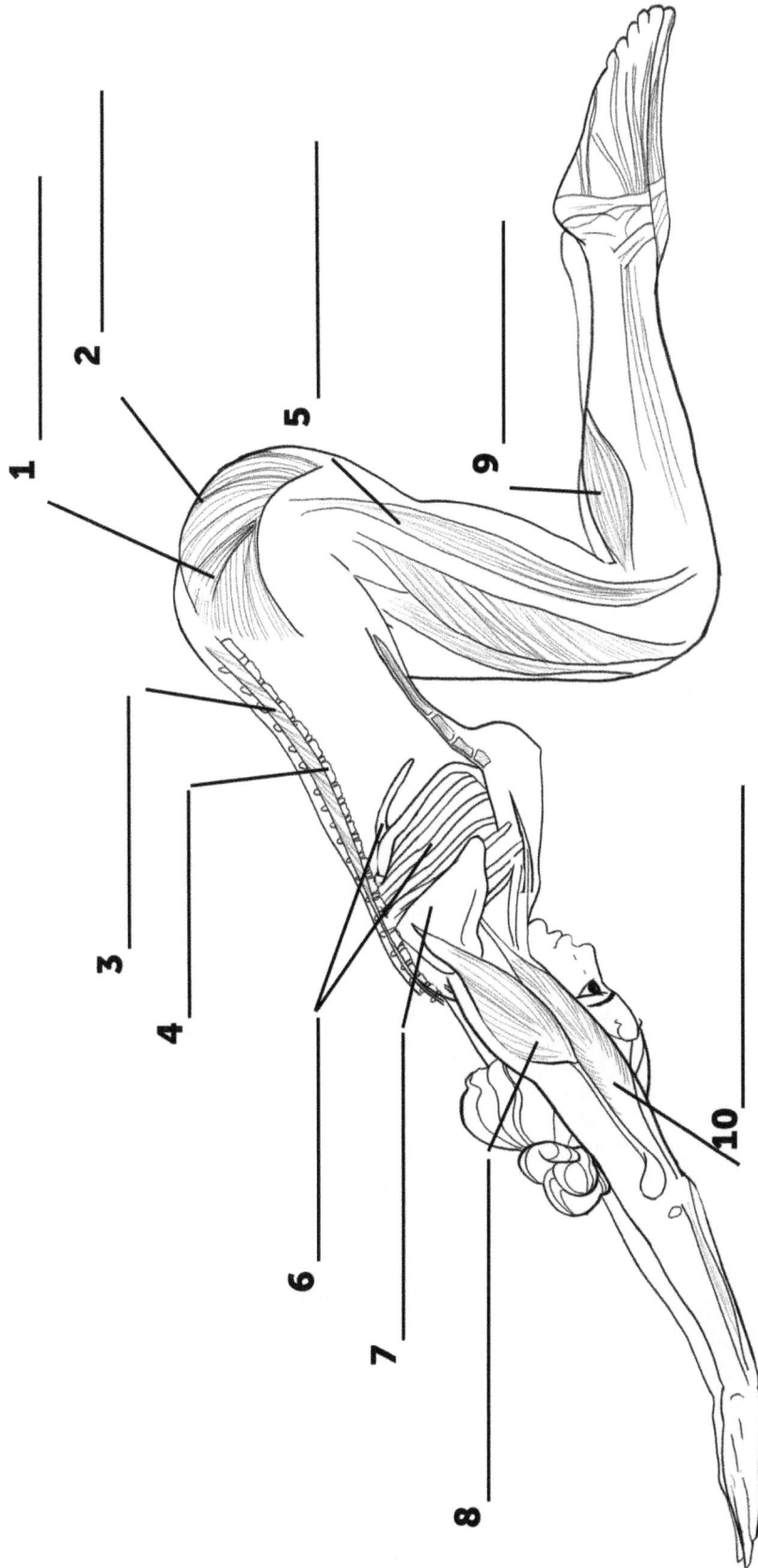

1

2

3

4

5

6

7

8

9

10

40. UTTANA SHISHOSANA

1. PIRIFORME
2. MÚSCULO GLÚTEO MAYOR
3. MÚSCULOS DE LA COLUMNA
4. COLUMNA VERTEBRAL
5. ISQUIOTIBIALES
6. COSTILLAS
7. ESCÁPULA
8. DELTOIDES
9. GASTROCNEMIO
10. TRÍCEPS BRAQUIAL

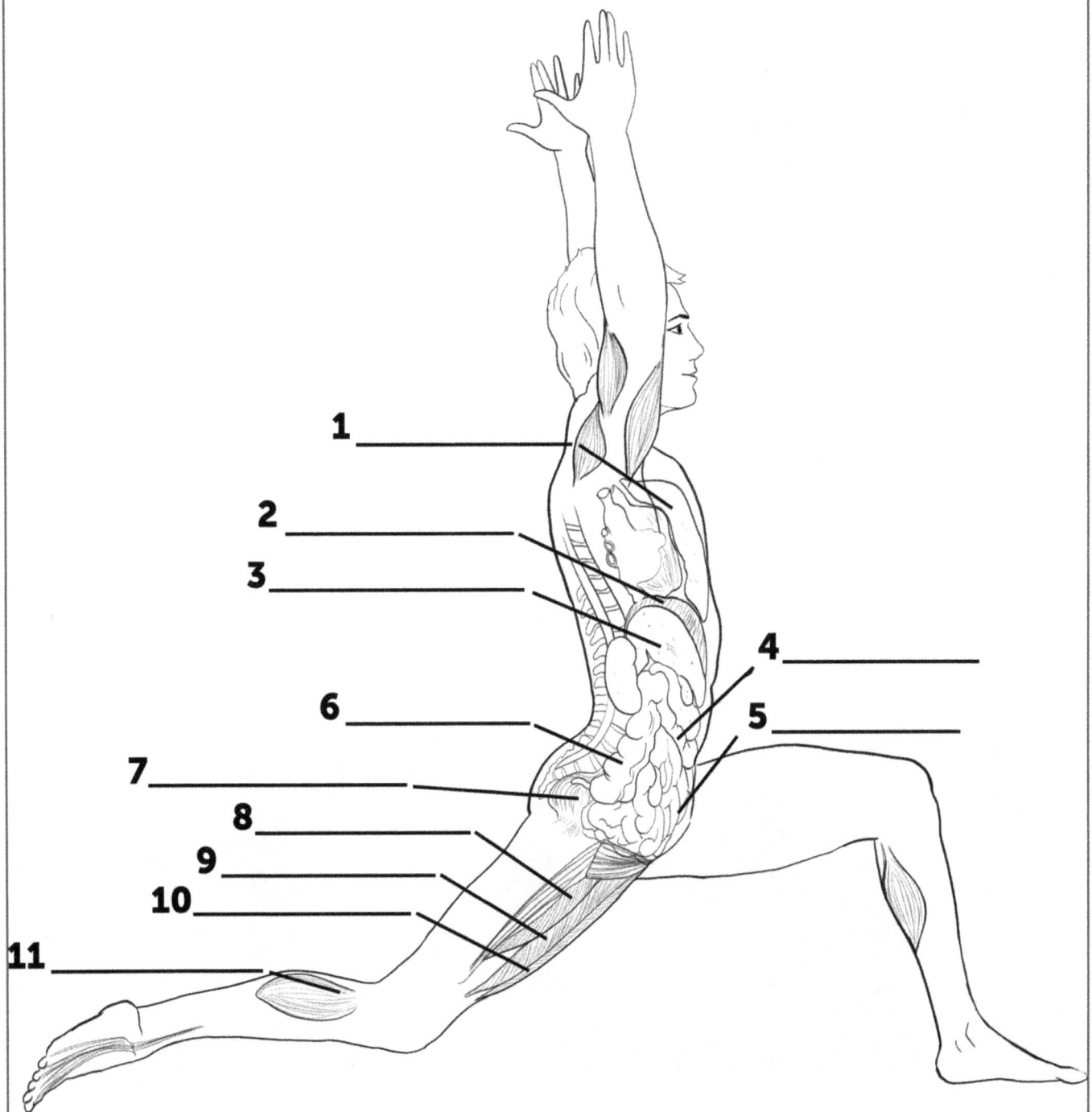

41. LUNGE BAJO

1 _____

2 _____

3 _____

4 _____

6 _____

5 _____

7 _____

8 _____

9 _____

10 _____

11 _____

41. LUNGE BAJO

1.   PULMONES

2.   DIAFRAGMA

3.   HÍGADO

4.   COLON TRANSVERSO

5.   FOLICULOS DE INTESTINO DELGADO

6.   COLON ASCENDENTE

7.   RECTO

8.   MÚSCULO VASTO LATERAL

9.   RECTO FEMORAL

10. VASTO MEDIAL

11. GASTROCNEMIO

42. PARVRTTA ANJANEYASANA

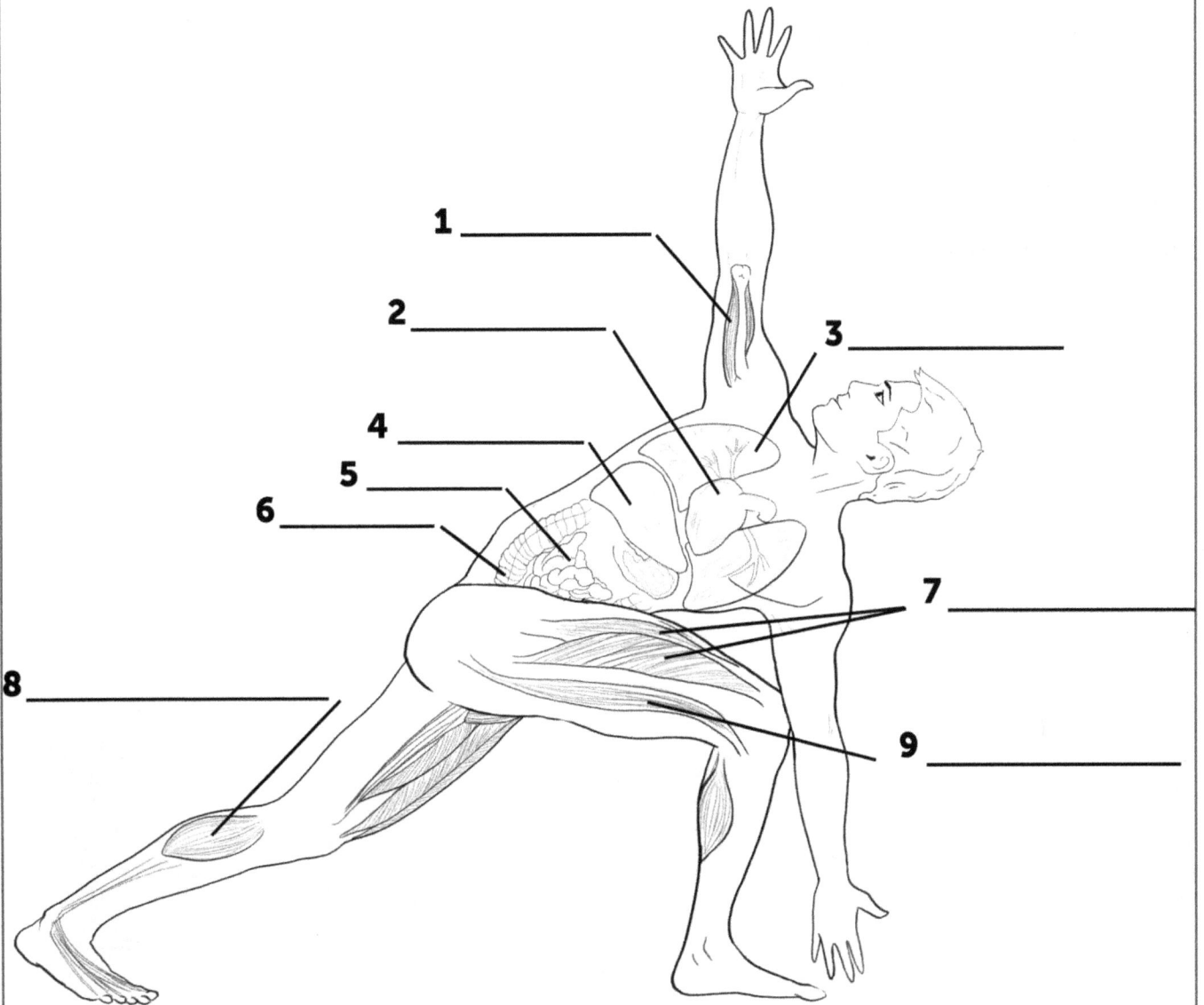

1 _____

2 _____

3 _____

4 _____

5 _____

6 _____

7 _____

8 _____

9 _____

42. PARVRTTA ANJANEYASANA

1.   BÍCEPS BRAQUIAL

2.   CORAZÓN

3.   PULMONES

4.   HÍGADO

5.   FOLICULOS DE INTESTINO DELGADO

6.   COLON ASCENDENTE

7.   CUADRÍCEPS

8.   GASTROCNEMIO

9.   ISQUIOTIBIALES

43. PRASARITA PADOTTANASANA

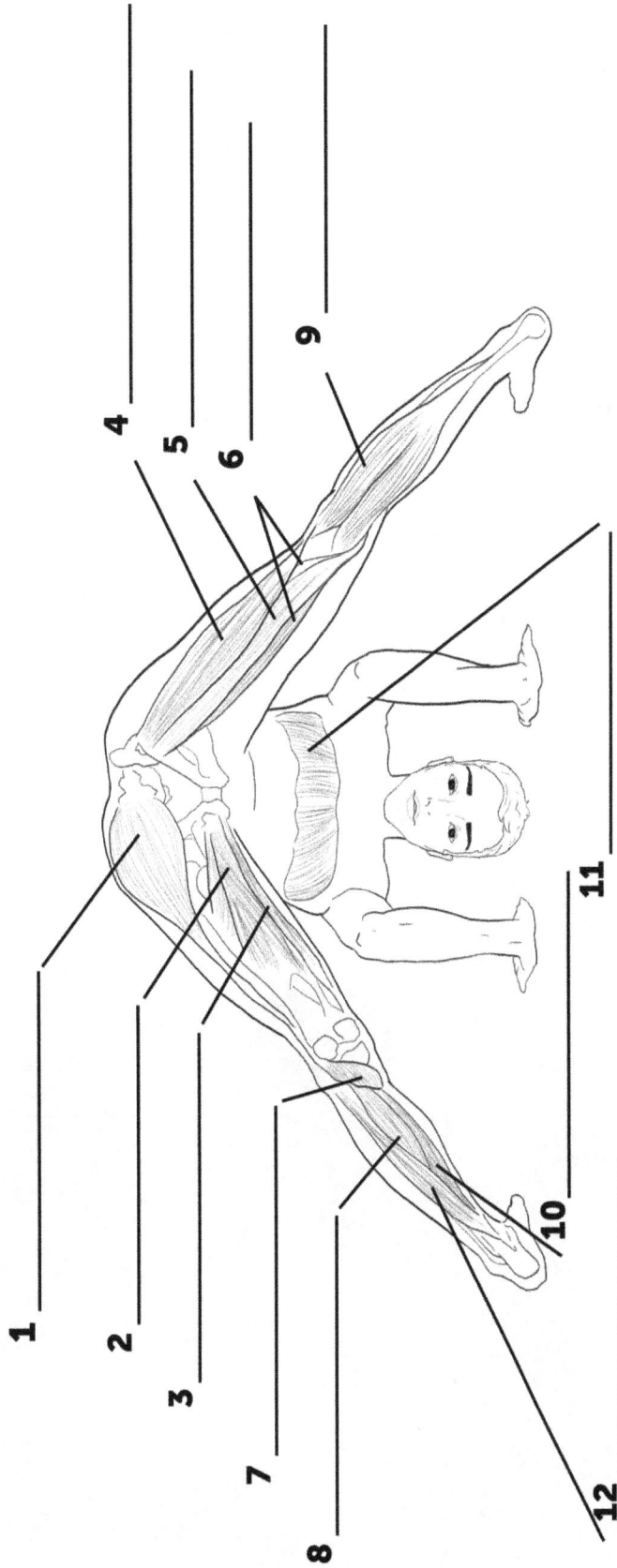

1

2

3

4

5

6

7

8

9

10

11

12

43. PRASARITA PADOTTANASANA

1. MÚSCULO GLÚTEO MAYOR
2. ADUCTOR MAYOR
3. GRÁCIL
4. BÍCEPS FEMORAL
5. SEMITENDINOSO
6. SEMIMEMBRANOSO
7. POPLÍTEO
8. TIBIAL POSTERIOR
9. GASTROCNEMIO
10. FLEXOR LARGO DE LOS DEDOS
11. DIAFRAGMA
12. FLEXOR LARGO DEL DEDO

44. POSTURA DE DIOSA

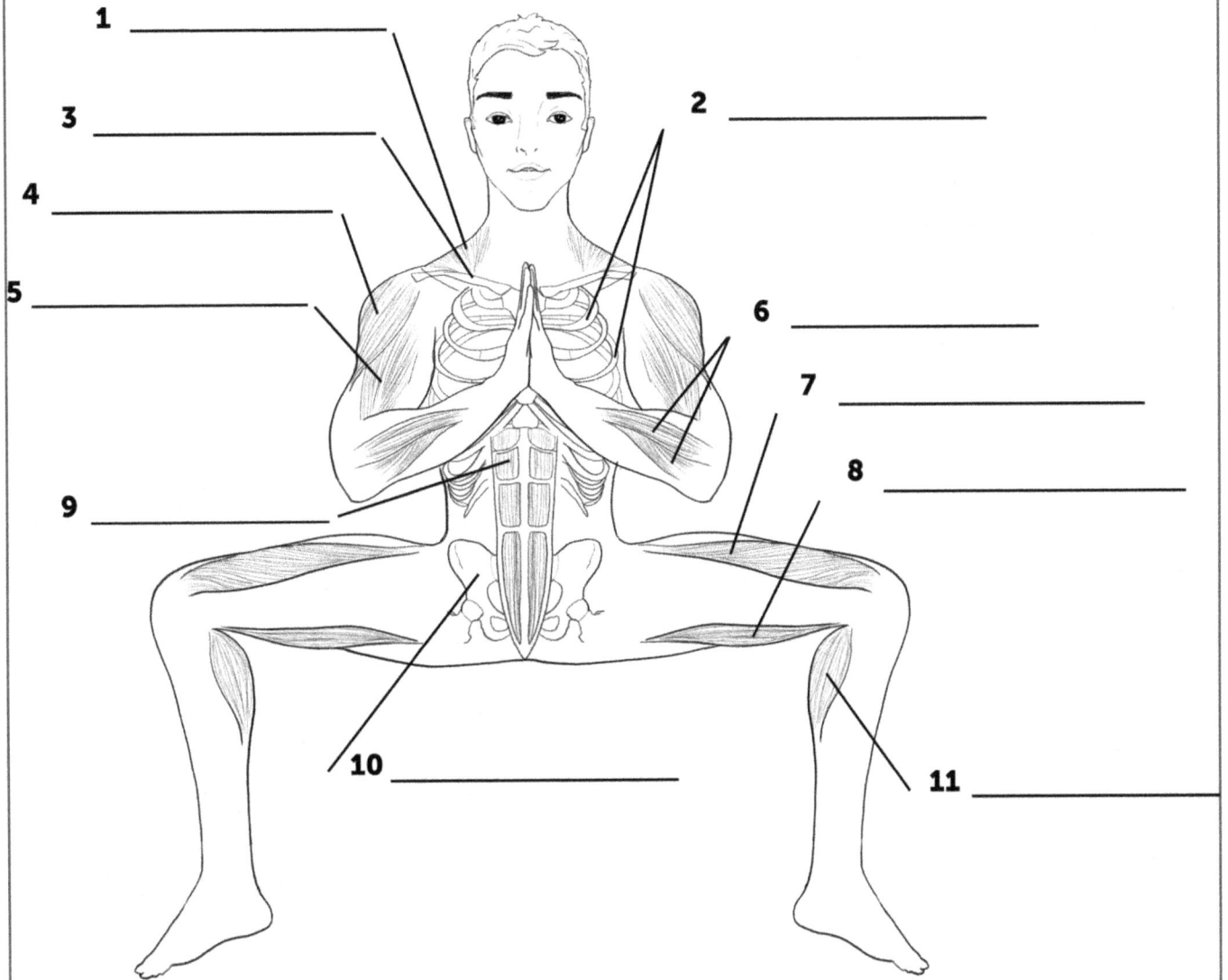

1 _____

3 _____

4 _____

5 _____

9 _____

2 _____

6 _____

7 _____

8 _____

10 _____

11 _____

44. POSTURA DE DIOSA

1. TRAPECIO
2. COSTILLAS
3. CLAVÍCULA
4. DELTOIDES
5. BÍCEPS BRAQUIAL
6. PRONADORES
7. CUADRÍCEPS
8. ISQUIOTIBIALES
9. RECTO ABDOMINAL
10. PELVIS
11. GASTROCNEMIO

45. PUENTE DE UNA PIERNA

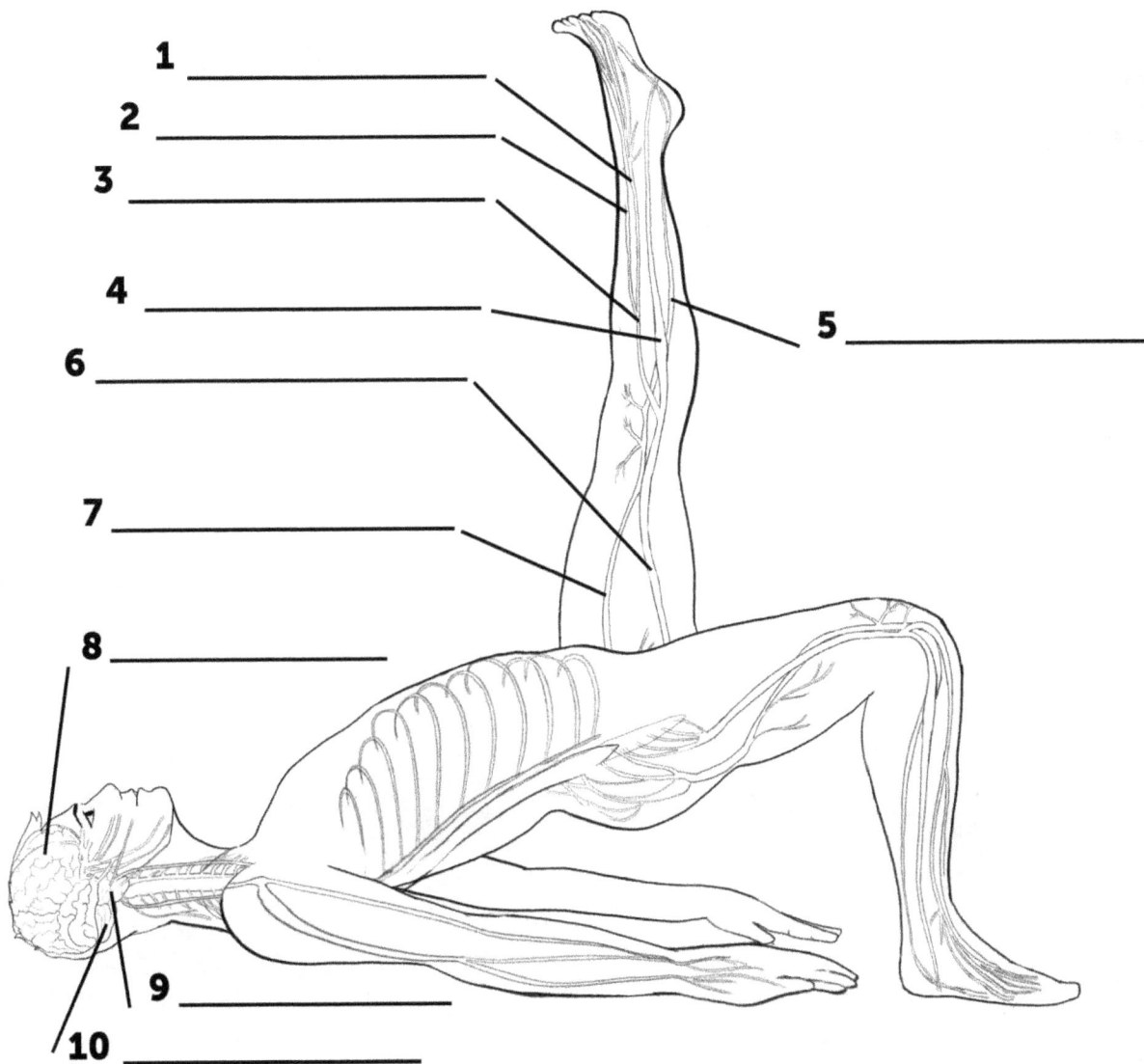

1 _____

2 _____

3 _____

4 _____

5 _____

6 _____

7 _____

8 _____

9 _____

10 _____

45. PUENTE DE UNA PIERNA

1. PERONEO PROFUNDO

2. PERONEO SUPERFICIAL

3. PERONEO COMÚN

4. TIBIAL

5. SAFENA

6. CIÁTICO

7. FEMORAL

8. CEREBRO

9. TRONCO ENCEFÁLICO

10. CEREBELO

46. POSTURA DEL CUERVO

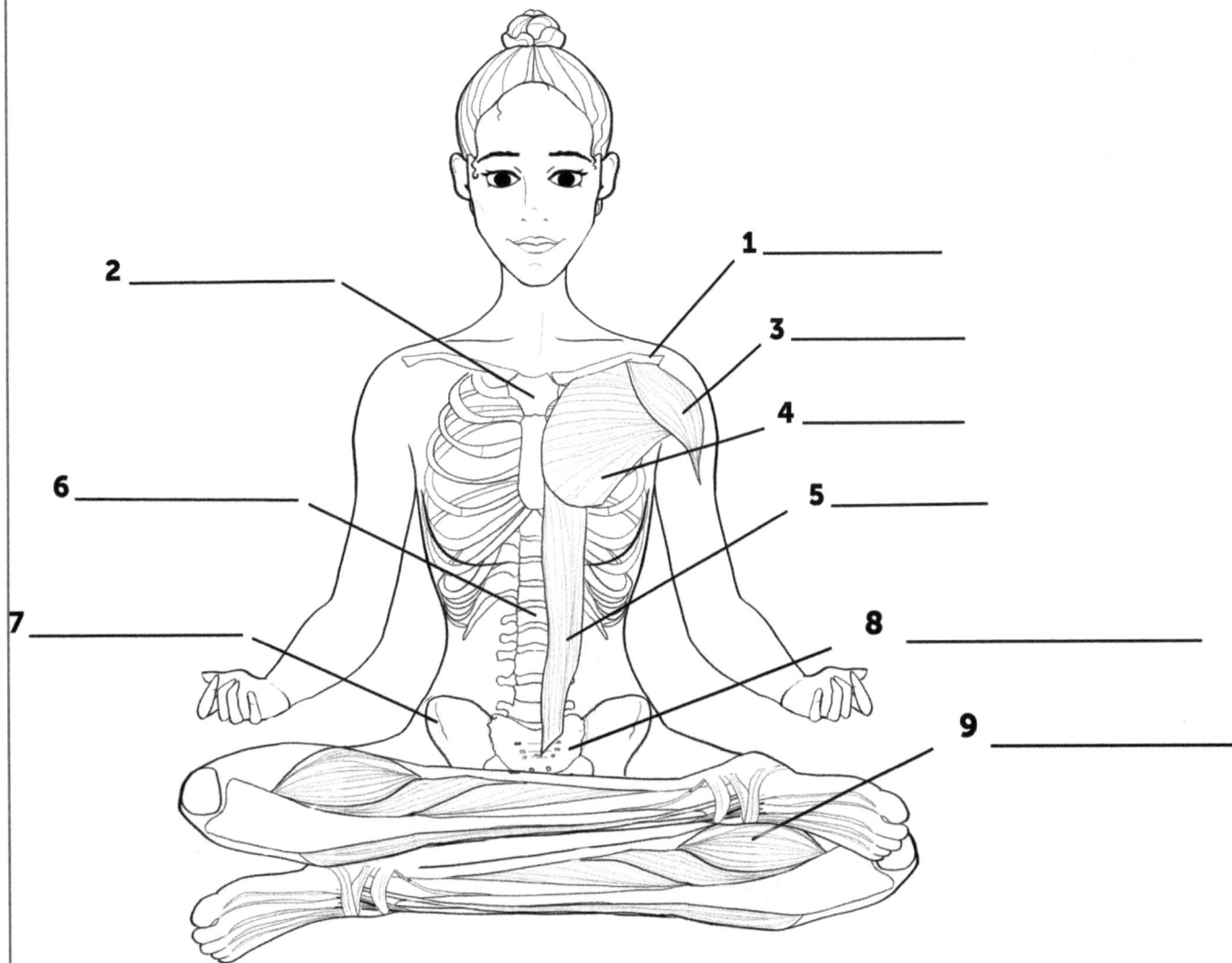

1 _____

2 _____

3 _____

4 _____

5 _____

6 _____

7 _____

8 _____

9 _____

46. POSTURA DEL CUERVO

1. CLAVÍCULA
2. ESTERNÓN
3. DELTOIDES
4. PECTORAL MAYOR
5. RECTO ABDOMINAL
6. COLUMNA VERTEBRAL
7. PELVIS
8. SACRO
9. GASTROCNEMIO

47. PASCHIMOTTANASANA

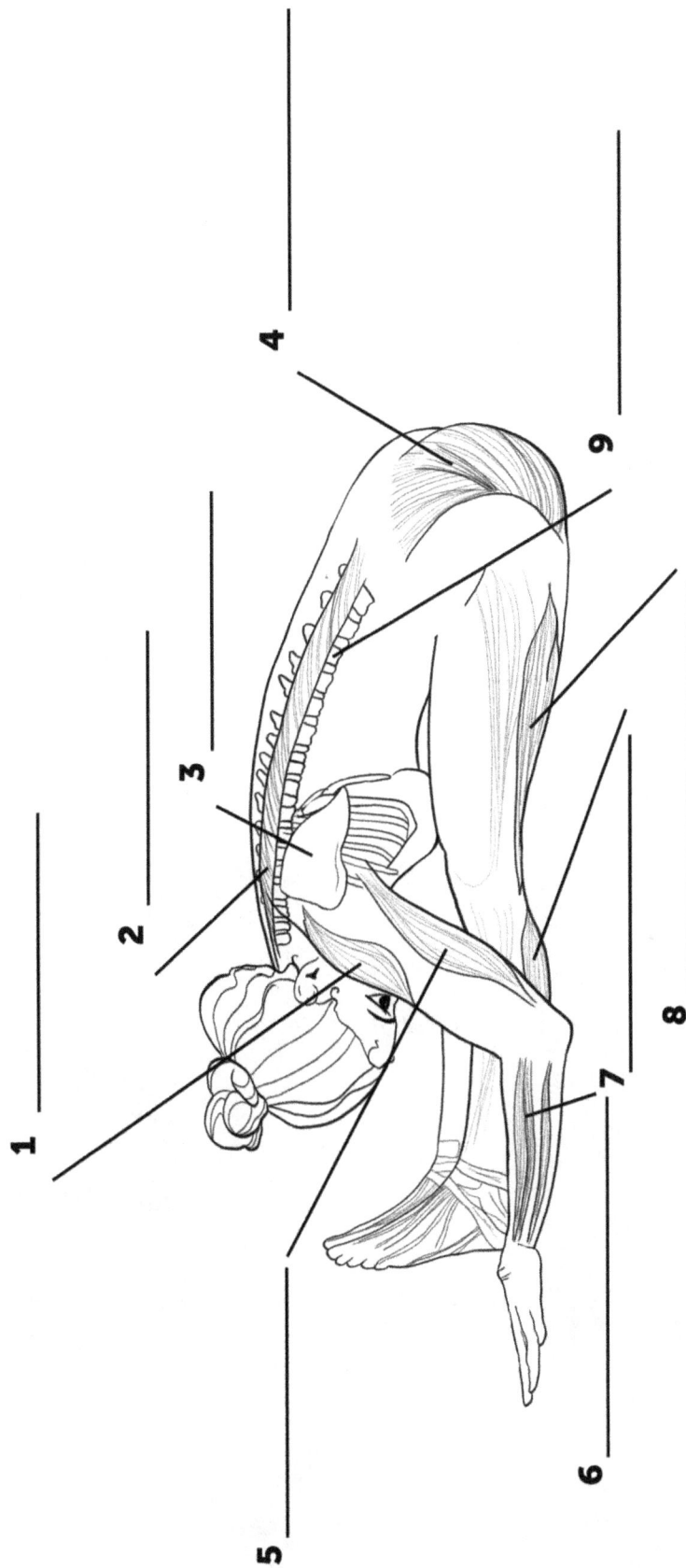

1

2

3

4

5

6

7

8

9

47. PASCHIMOTTANASANA

1. DELTOIDES
2. MÚSCULOS DE LA COLUMNA
3. ESCÁPULA
4. PIRIFORME
5. TRÍCEPS BRAQUIAL
6. PRONADORES
7. GASTROCNEMIO
8. ISQUIOTIBIALES
9. COLUMNA VERTEBRAL

48. JANU SHIRASASANA

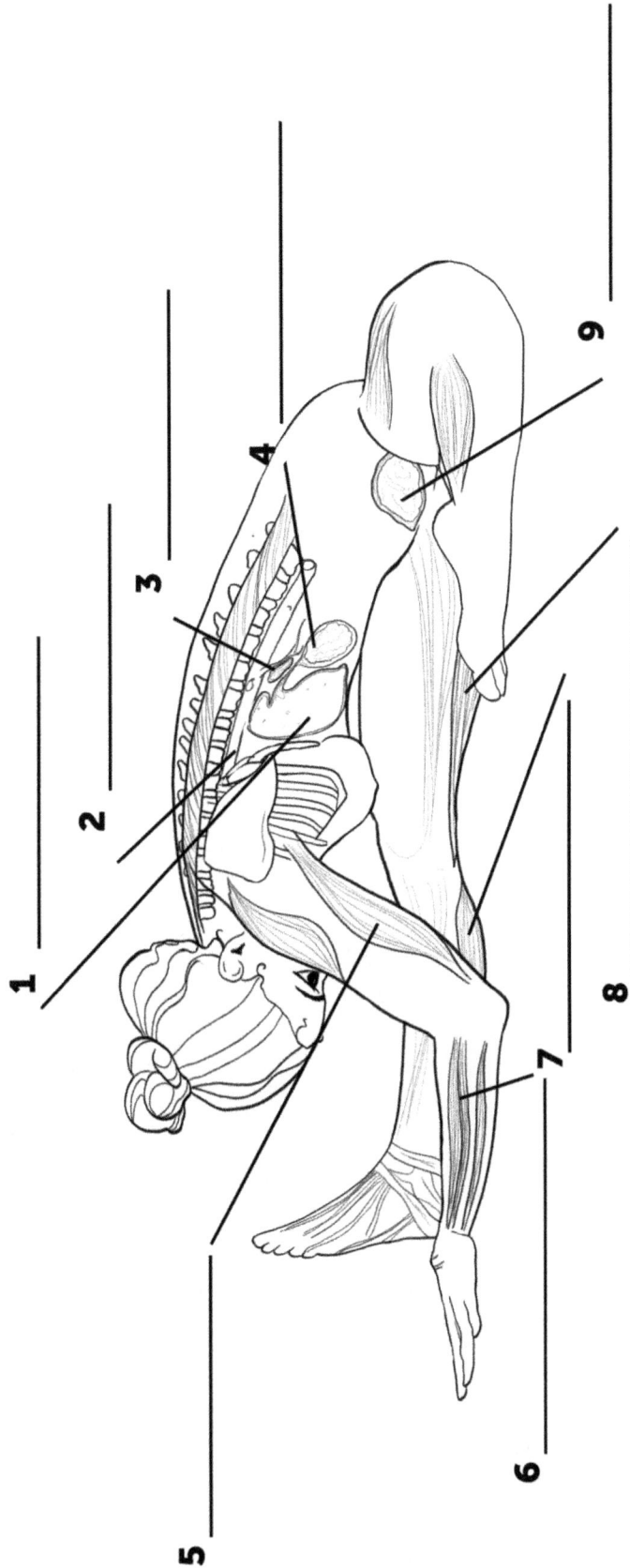

48. JANU SHIRASASANA

1. HÍGADO
2. AORTA ABDOMINAL
3. PÁNCREAS
4. ESTÓMAGO
5. TRÍCEPS BRAQUIAL
6. PRONADORES
7. GASTROCNEMIO
8. ISQUIOTIBIALES
9. VEJIGA URINARIA

49. RODILLAS AL PECHO

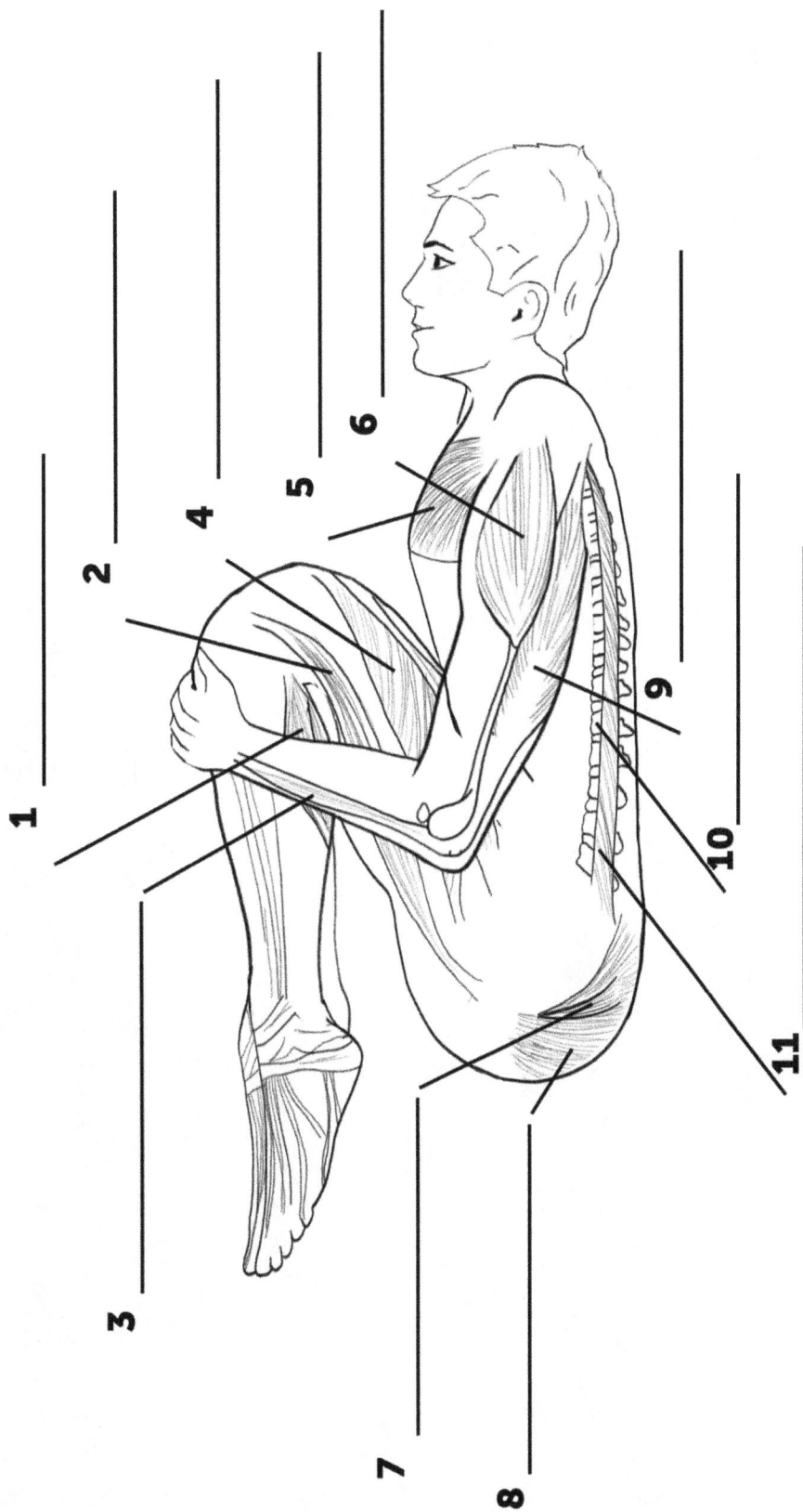

49. RODILLAS AL PECHO

1. GASTROCNEMIO
2. ISQUIOTIBIALES
3. PRONADORES
4. CUADRÍCEPS
5. PECTORAL MAYOR
6. DELTOIDES
7. PIRIFORME
8. MÚSCULO GLÚTEO MAYOR
9. TRÍCEPS BRAQUIAL
10. COLUMNA VERTEBRAL
11. MÚSCULOS DE LA COLUMNA

50. POSTURA DE LEÓN

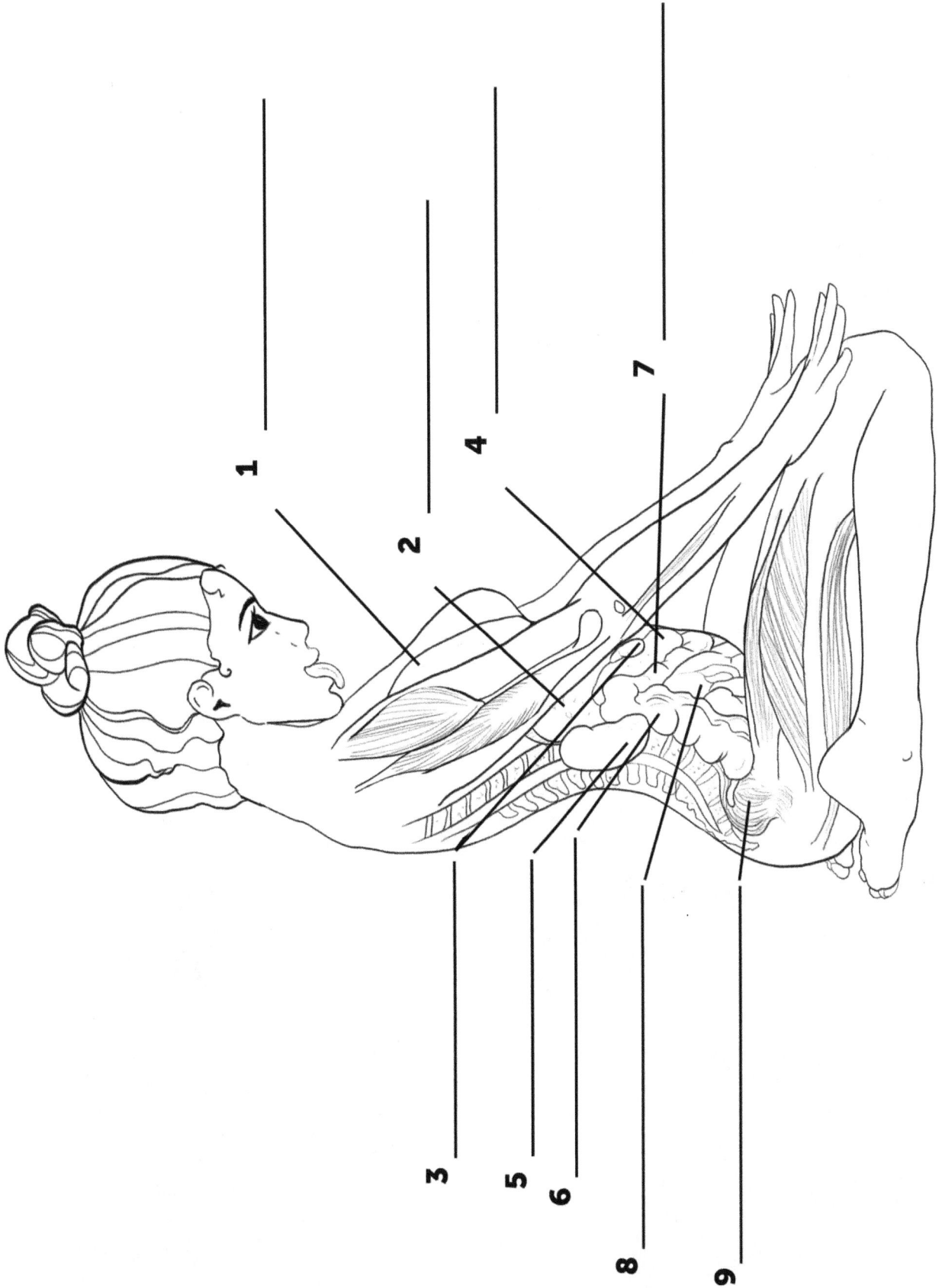

1

2

4

7

3

5

6

8

9

50. POSTURA DE LEÓN

1. PULMONES

2. HÍGADO

3. VESÍCULA BILIAR

4. ESTÓMAGO

5. RIÑÓN

6. COLON ASCENDENTE

7. COLON TRANSVERSO

8. FOLICULOS DE INTESTINO DELGADO

9. RECTO

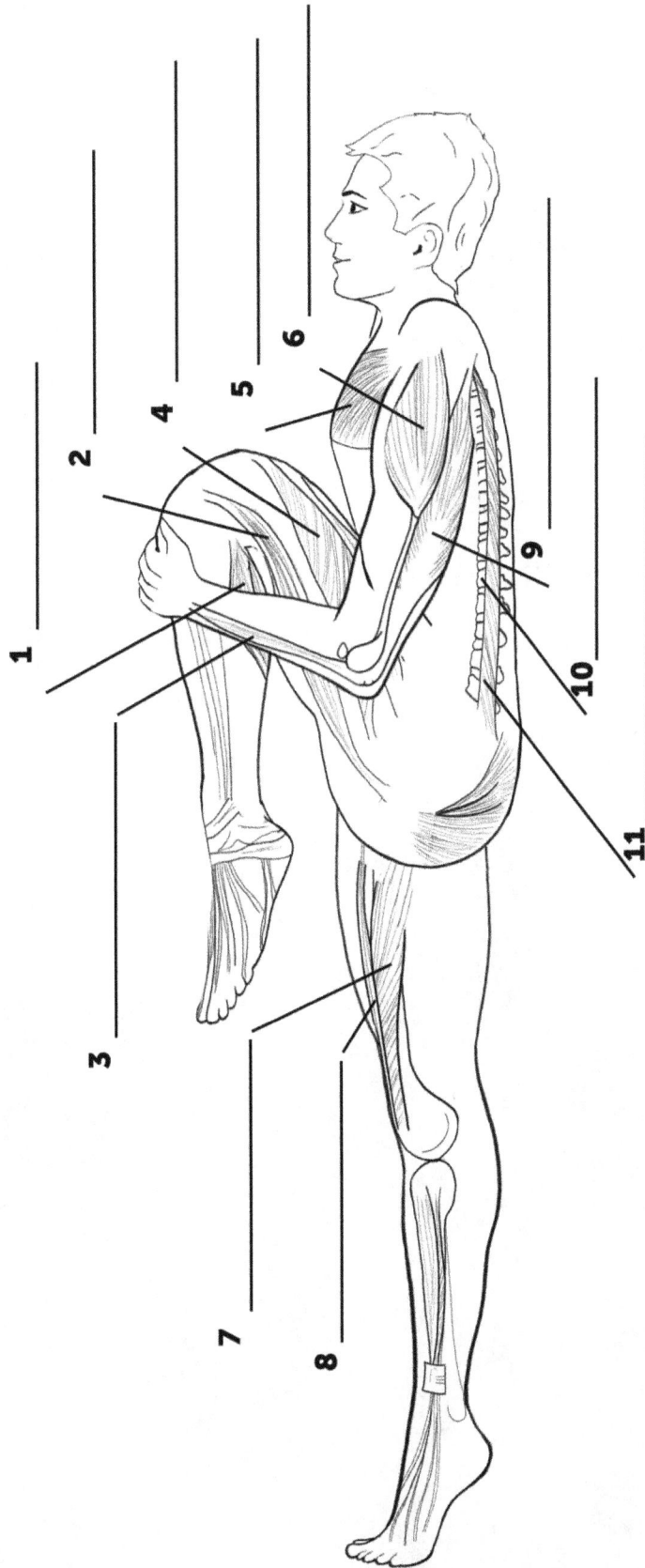

51. SAVASANA

51. SAVASANA

1. GASTROCNEMIO
2. ISQUIOTIBIALES
3. PRONADORES
4. CUADRÍCEPS
5. PECTORAL MAYOR
6. DELTOIDES
7. RECTO FEMORAL
8. SARTORIO
9. TRÍCEPS BRAQUIAL
10. COLUMNA VERTEBRAL
11. MÚSCULOS DE LA COLUMNA

52. GATO SENTADO

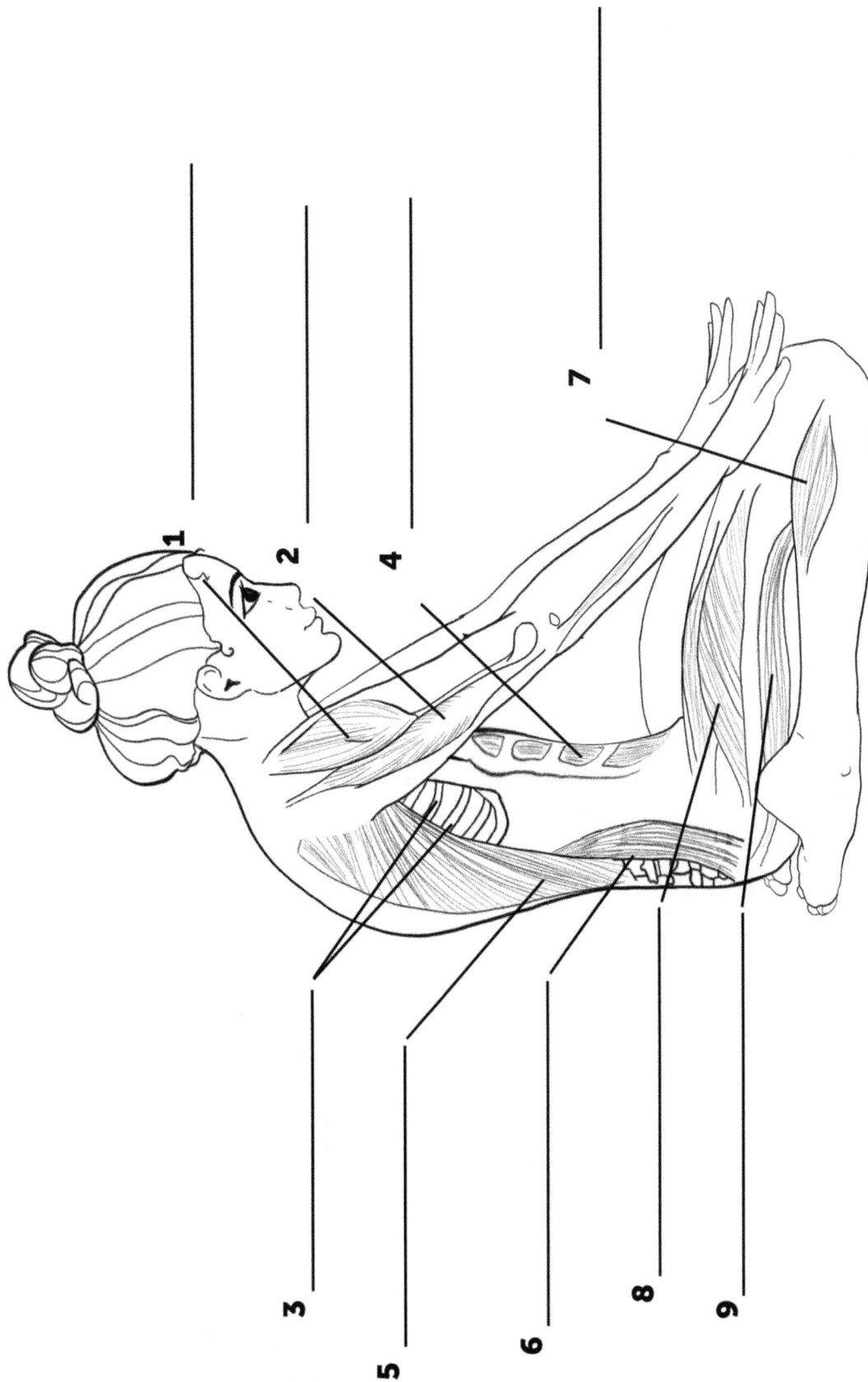

52. GATO SENTADO

1. DELTOIDES
2. TRÍCEPS BRAQUIAL
3. COSTILLAS
4. RECTO ABDOMINAL
5. LATISSIMUS DORSI
6. ERECTOR DE LA COLUMNA
7. GASTROCNEMIO
8. CUADRÍCEPS
9. ISQUIOTIBIALES

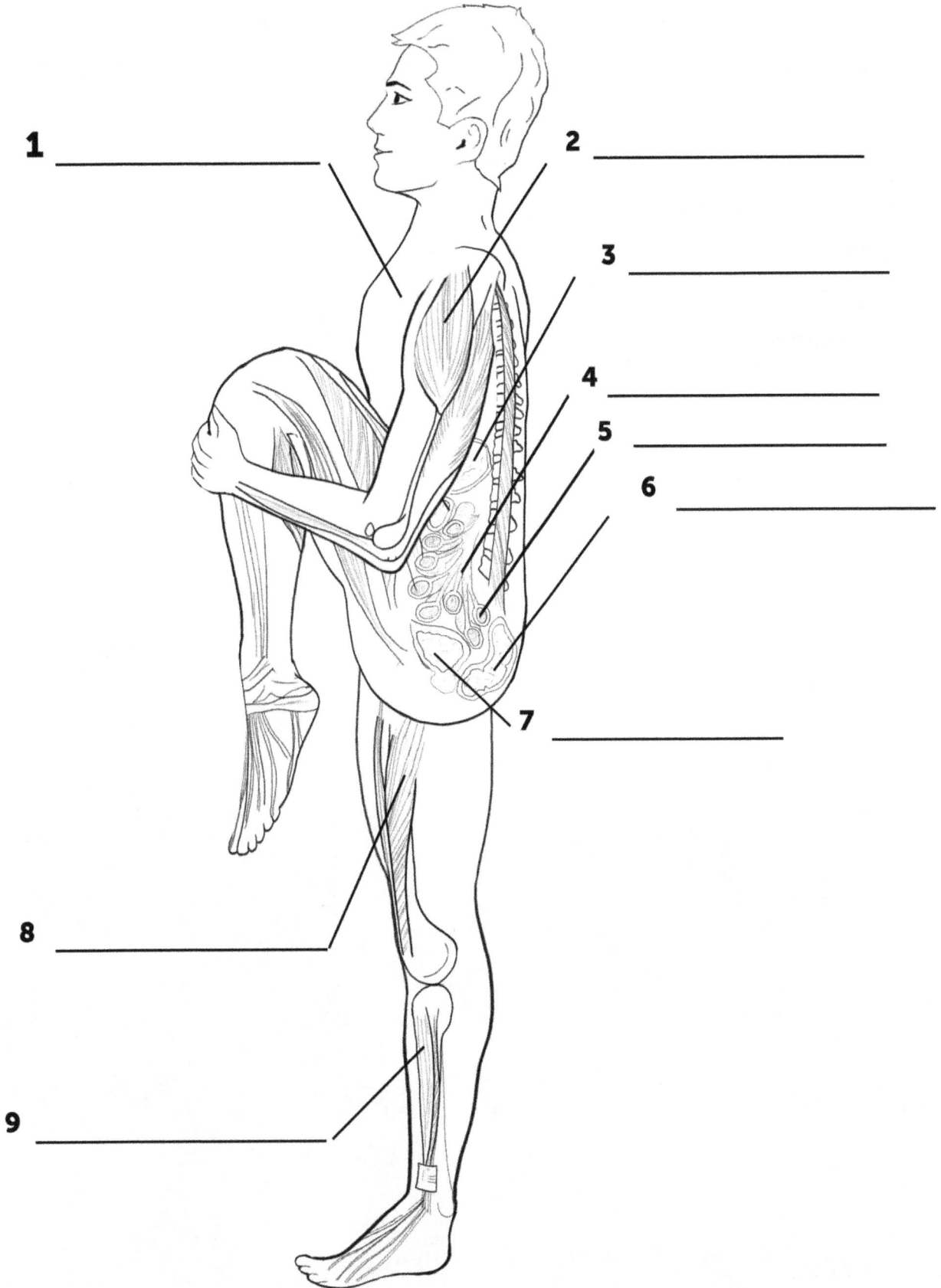

53. VṛKṣāSANA

1 _____

2 _____

3 _____

4 _____

5 _____

6 _____

7 _____

8 _____

9 _____

53. Vṛkṣāsana

1. PECHO
2. DELTOIDES
3. ESTÓMAGO
4. MESENTERIO DEL INTESTINO DELGADO
5. FOLICULOS DE INTESTINO DELGADO
6. RECTO
7. VEJIGA URINARIA
8. RECTO FEMORAL
9. TIBIAL ANTERIOR

54. POSTURA DE MEDIO LOTO DE PIE

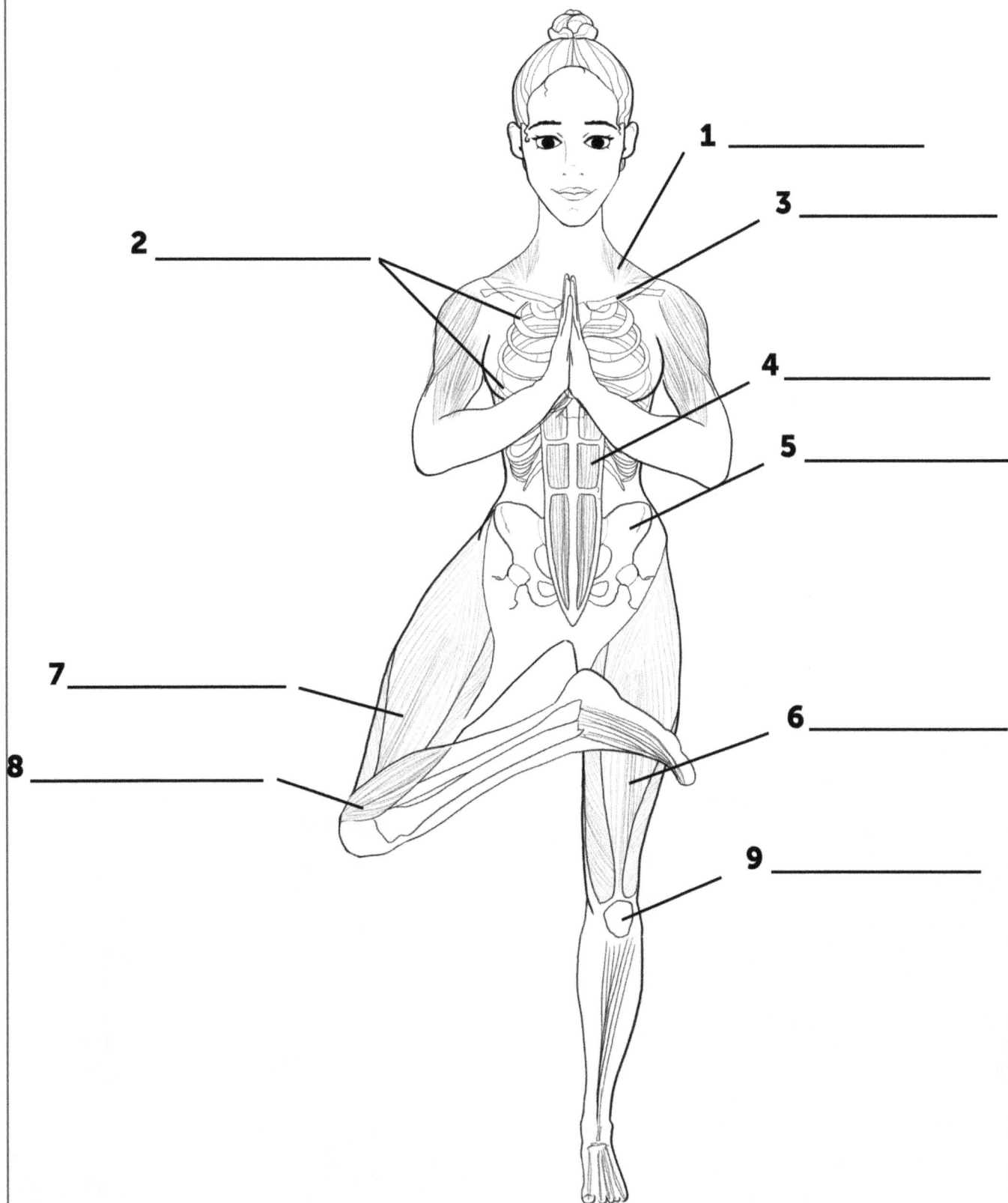

1 _____

2 _____

3 _____

4 _____

5 _____

6 _____

7 _____

8 _____

9 _____

54. POSTURA DE MEDIO LOTO DE PIE

1. TRAPECIO

2. COSTILLAS

3. CLAVÍCULA

4. RECTO ABDOMINAL

5. PELVIS

6. CUADRÍCEPS

7. ISQUIOTIBIALES

8. GASTROCNEMIO

9. RÓTULA

POSTURAS DE YOGA PARA INTERMEDIOS

55. VASISTHASANA

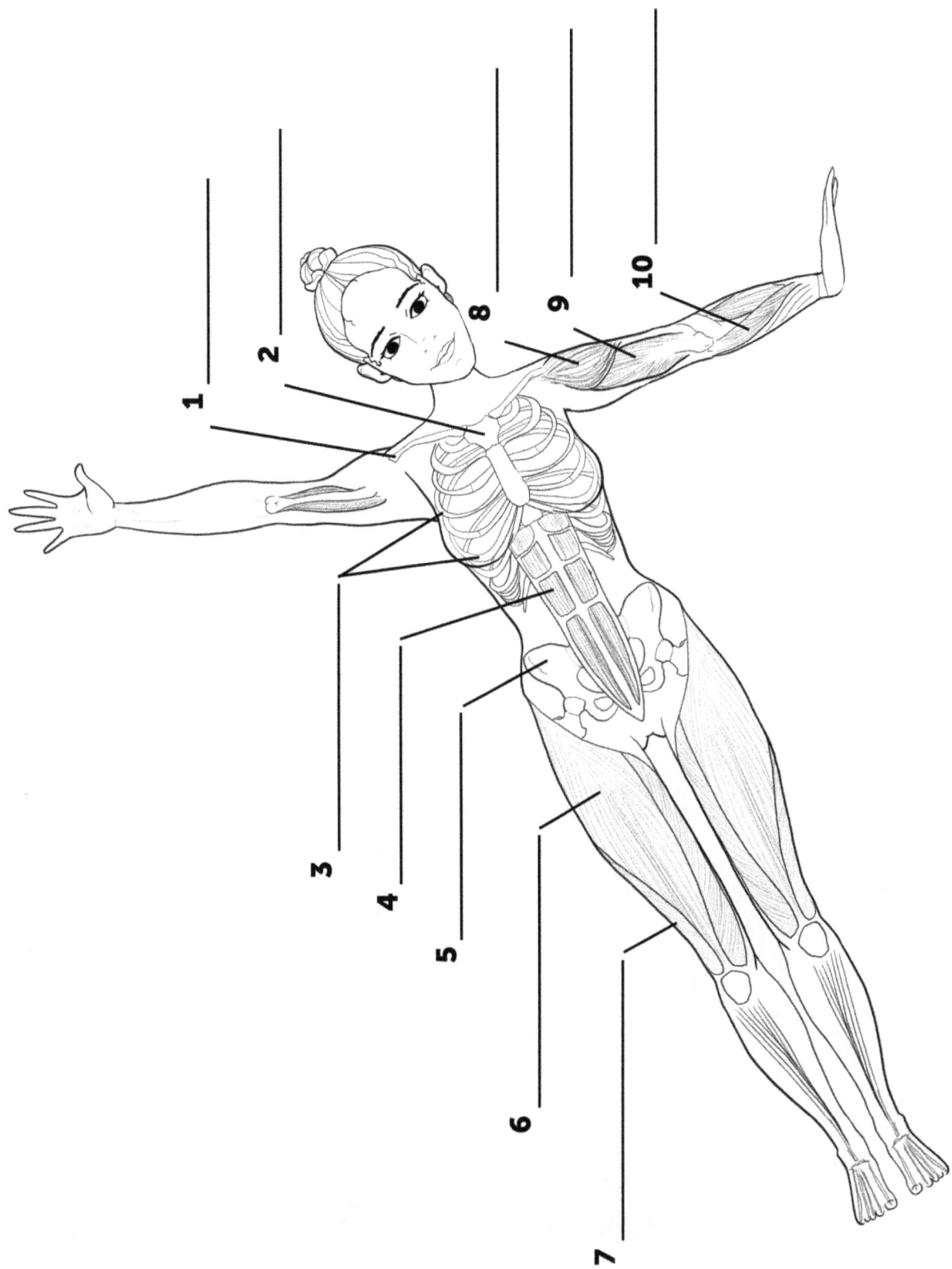

55. VASISTHASANA

1. CLAVÍCULA
2. ESTERNÓN
3. COSTILLAS
4. RECTO ABDOMINAL
5. PELVIS
6. CUADRÍCEPS
7. MÚSCULO VASTO LATERAL
8. DELTOIDES
9. BÍCEPS BRAQUIAL
10. PRONADORES

56. CAMATKARASANA

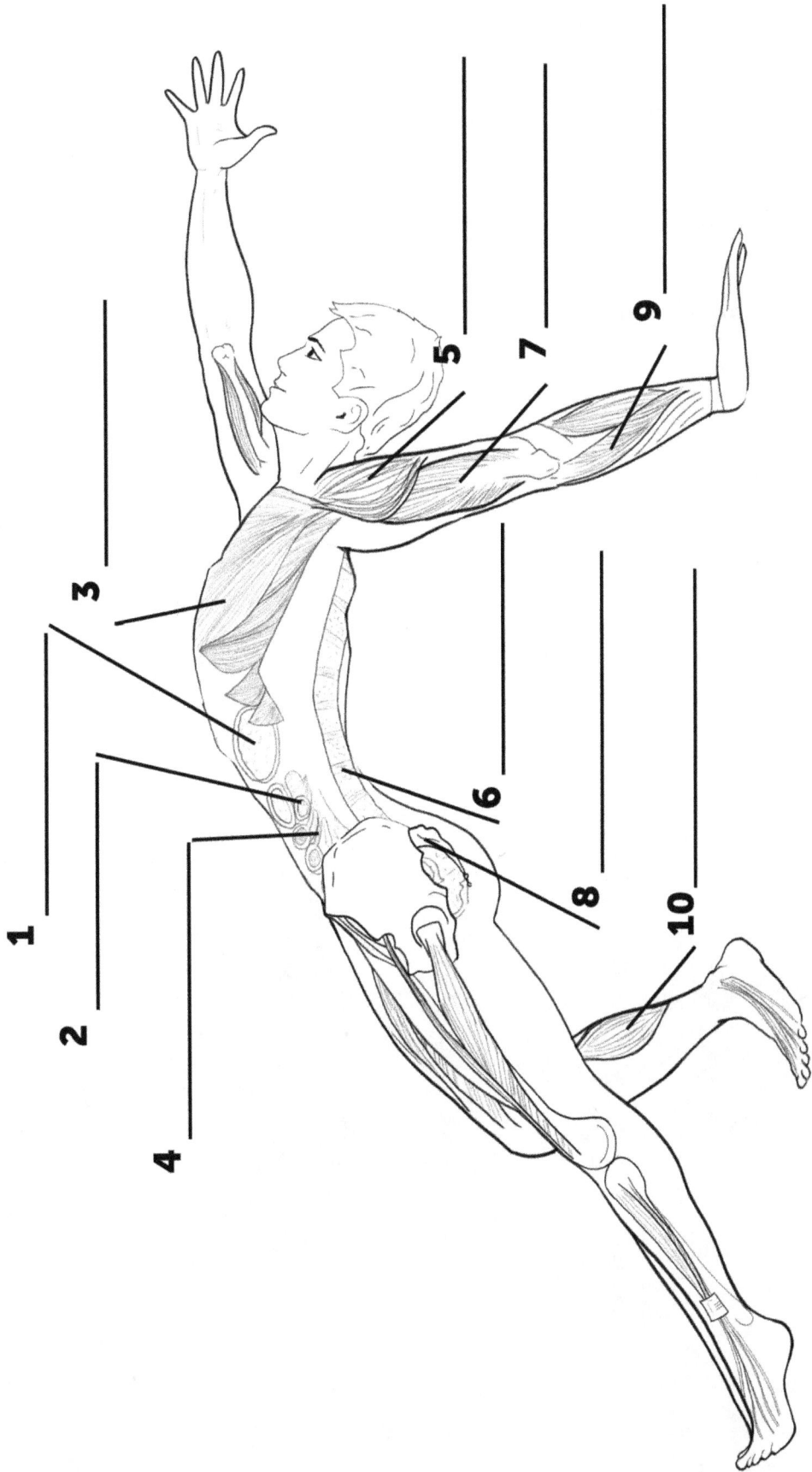

56. CAMATKARASANA

1. ESTÓMAGO
2. FOLICULOS DE INTESTINO DELGADO
3. PECTORAL MAYOR
4. MESENTERIO DEL INTESTINO DELGADO
5. DELTOIDES
6. COLUMNA VERTEBRAL
7. BÍCEPS BRAQUIAL
8. SACRO
9. PRONADORES
10. GASTROCNEMIO

57. POSTURA DE MEDIA RANA

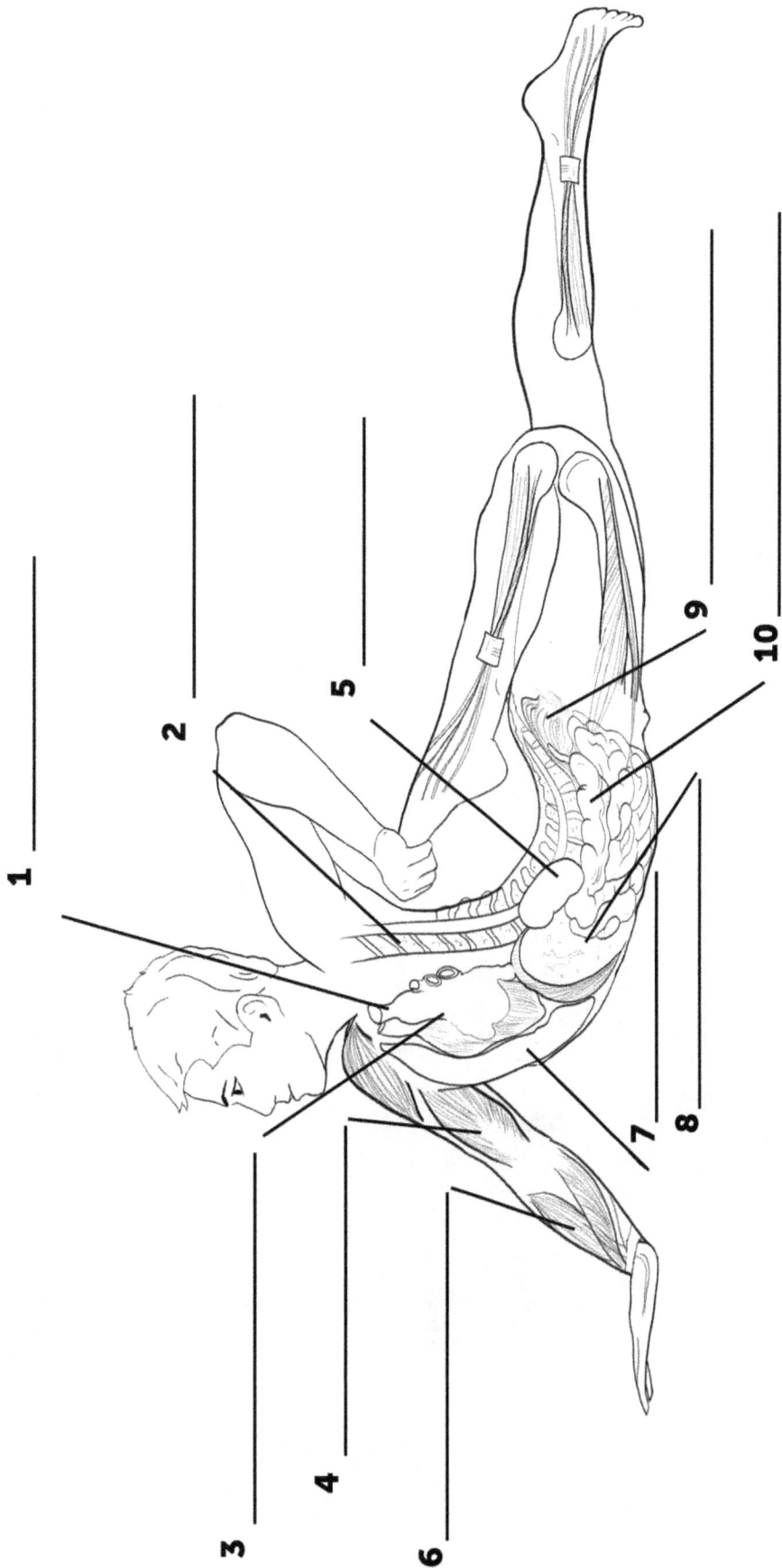

57. POSTURA DE MEDIA RANA

1.   AORTA

2.   COLUMNA VERTEBRAL

3.   CORAZÓN

4.   BÍCEPS BRAQUIAL

5.   RIÑÓN

6.   PRONADORES

7.   PULMONES

8.   HÍGADO

9.   RECTO

10.   COLON ASCENDENTE

58. PARIVRTTA SURYA YANTRASANA

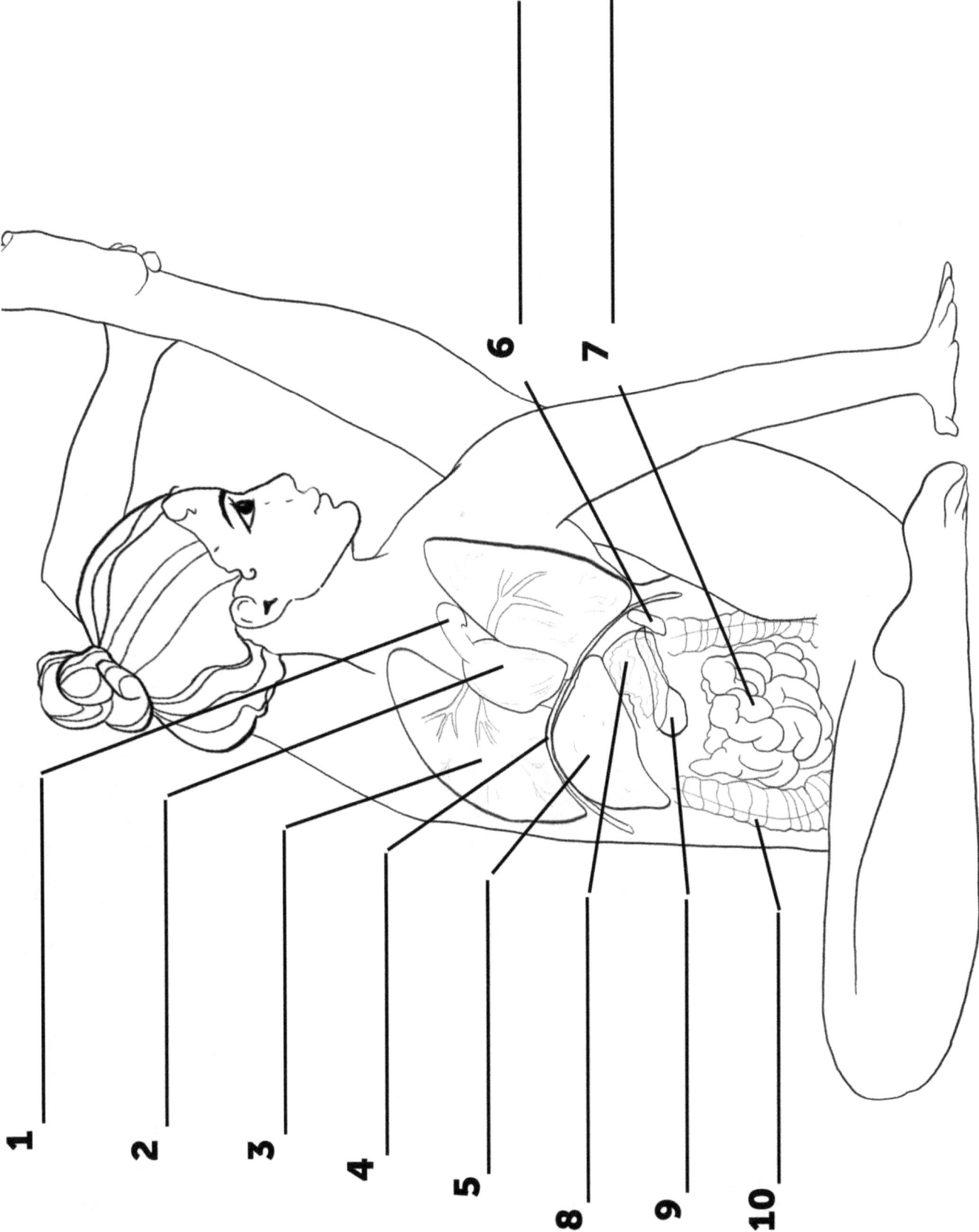

1

2

3

4

5

6

7

8

9

10

58. PARIVRTTA SURYA YANTRASANA

1. AORTA
2. CORAZÓN
3. PULMONES
4. DIAFRAGMA
5. HÍGADO
6. VESÍCULA BILIAR
7. FOLICULOS DE INTESTINO DELGADO
8. ESTÓMAGO
9. PÁNCREAS
10. COLON ASCENDENTE

59. POSTURA MARICHI I

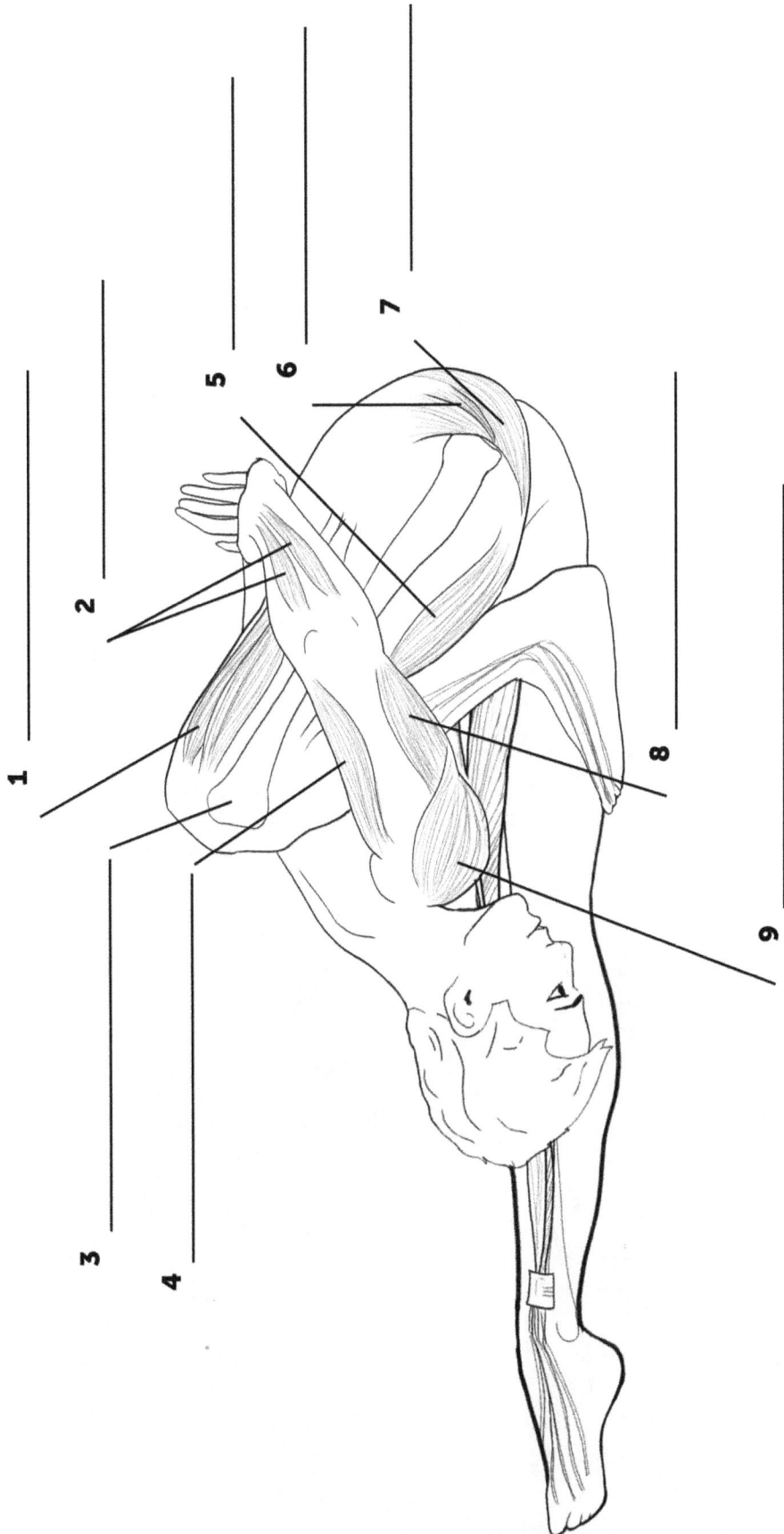

59. POSTURA MARICHI I

1. CUADRÍCEPS
2. PRONADORES
3. FÉMUR
4. BÍCEPS BRAQUIAL
5. ISQUIOTIBIALES
6. PIRIFORME
7. MÚSCULO GLÚTEO MAYOR
8. TRÍCEPS BRAQUIAL
9. DELTOIDES

60. POSTURA MARICHI II

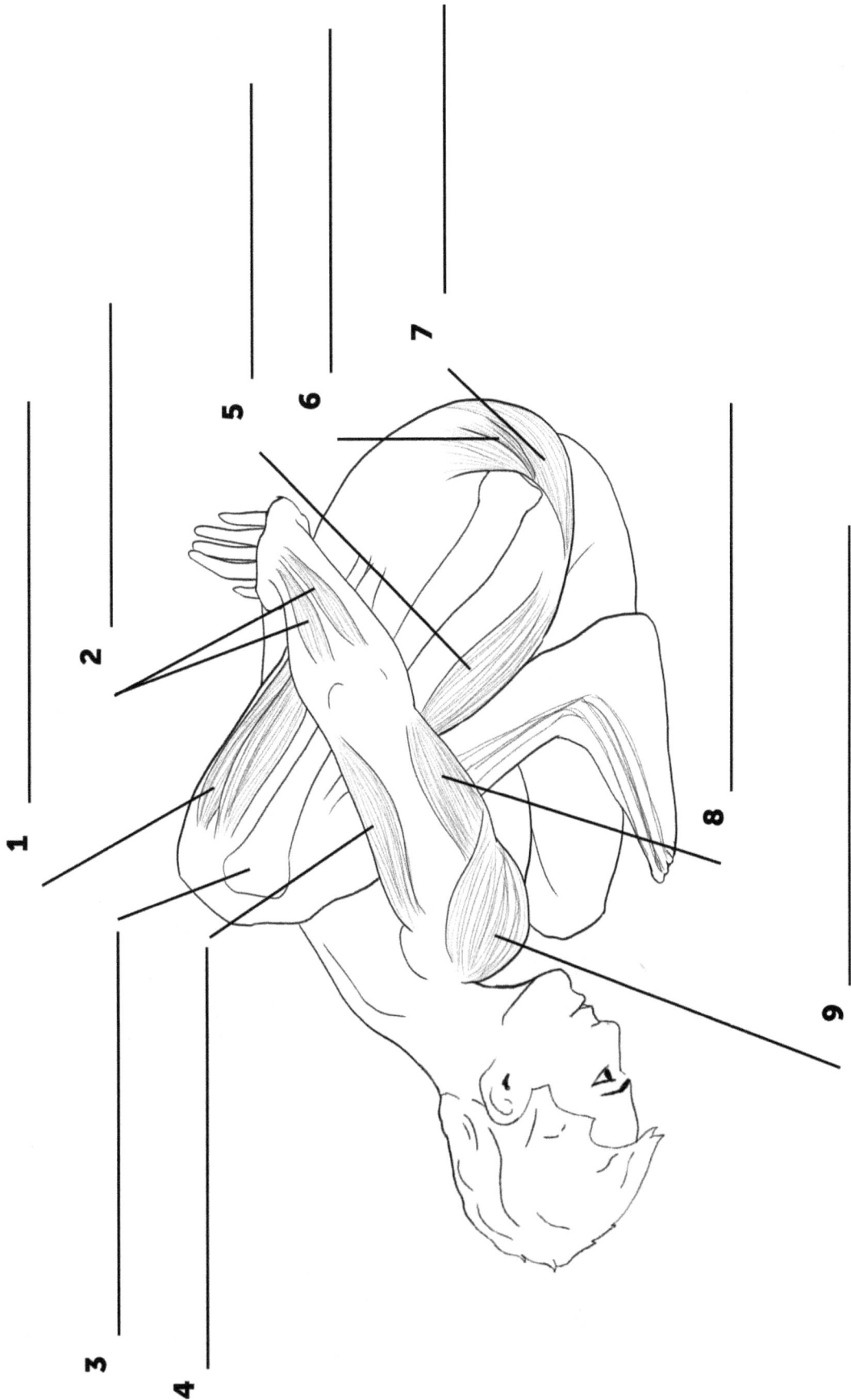

1

2

3

4

5

6

7

8

9

60. POSTURA MARICHI II

1. CUADRÍCEPS
2. PRONADORES
3. FÉMUR
4. BÍCEPS BRAQUIAL
5. ISQUIOTIBIALES
6. PIRIFORME
7. MÚSCULO GLÚTEO MAYOR
8. TRÍCEPS BRAQUIAL
9. DELTOIDES

61. POSTURA MARICHI II

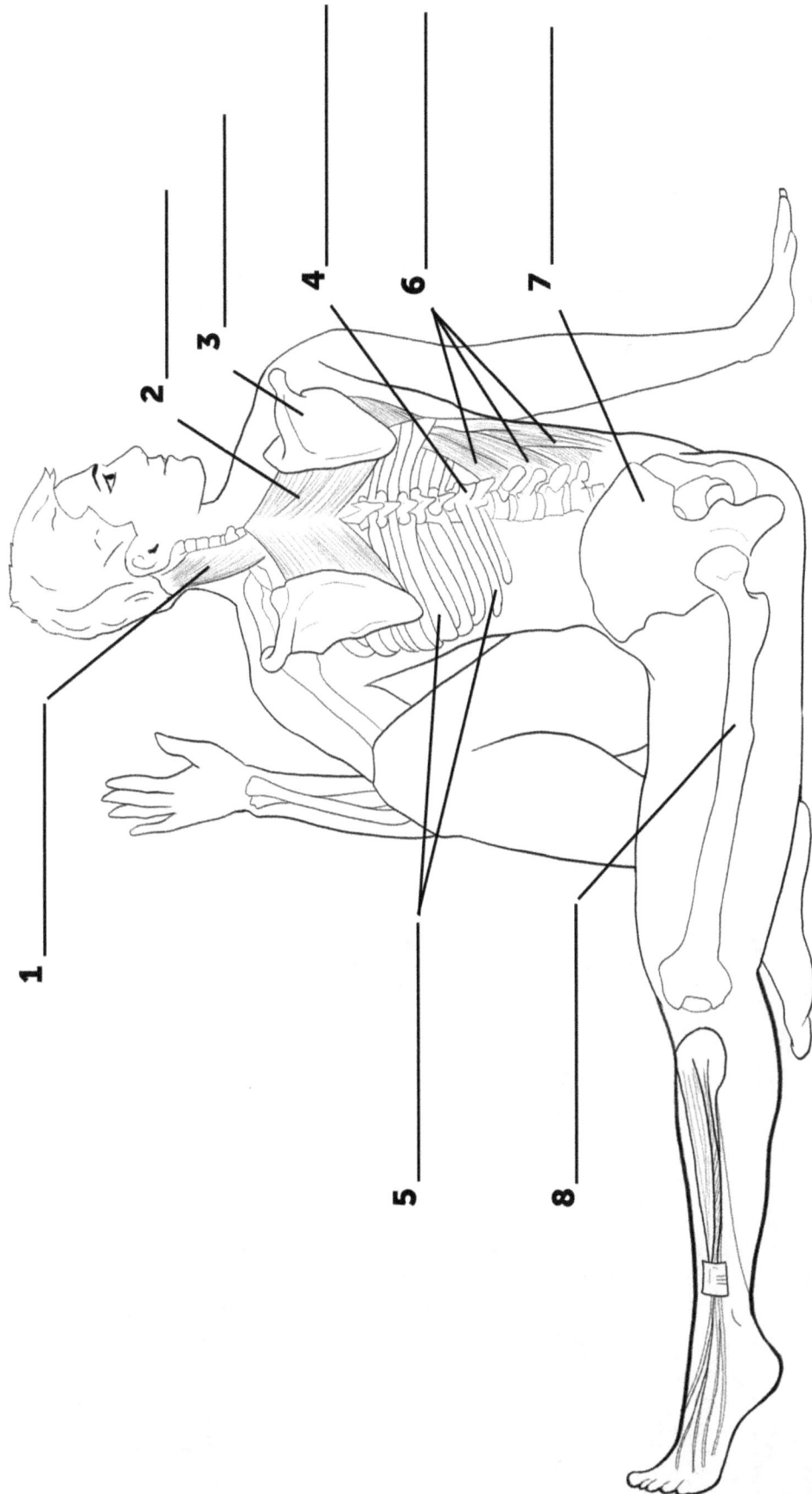

61. POSTURA MARICHI II

1. MÚSCULO ESPLENIO DE LA CABEZA
2. ROMBOIDES
3. ESCÁPULA
4. COLUMNA VERTEBRAL
5. COSTILLAS
6. ERECTOR DE LA COLUMNA
7. PELVIS
8. FÉMUR

62. POSTURA PIRÁMIDE

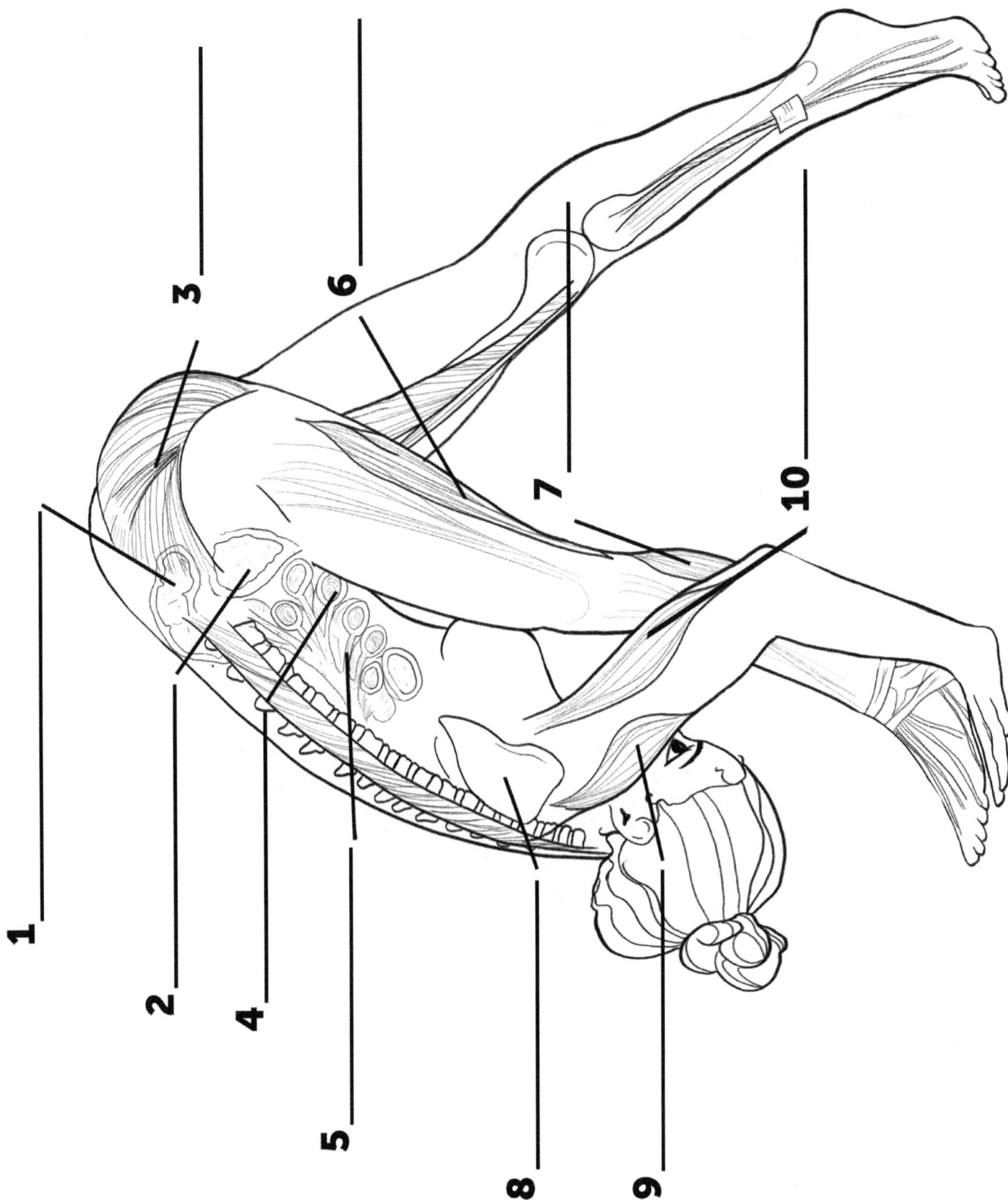

62. POSTURA PIRÁMIDE

1. RECTO

2. VEJIGA URINARIA

3. PIRIFORME

4. FOLICULOS DE INTESTINO DELGADO

5. MESENTERIO DEL INTESTINO DELGADO

6. ISQUIOTIBIALES

7. GASTROCNEMIO

8. ESCÁPULA

9. DELTOIDES

10. TRÍCEPS BRAQUIAL

63. VIRABHADRASANA I

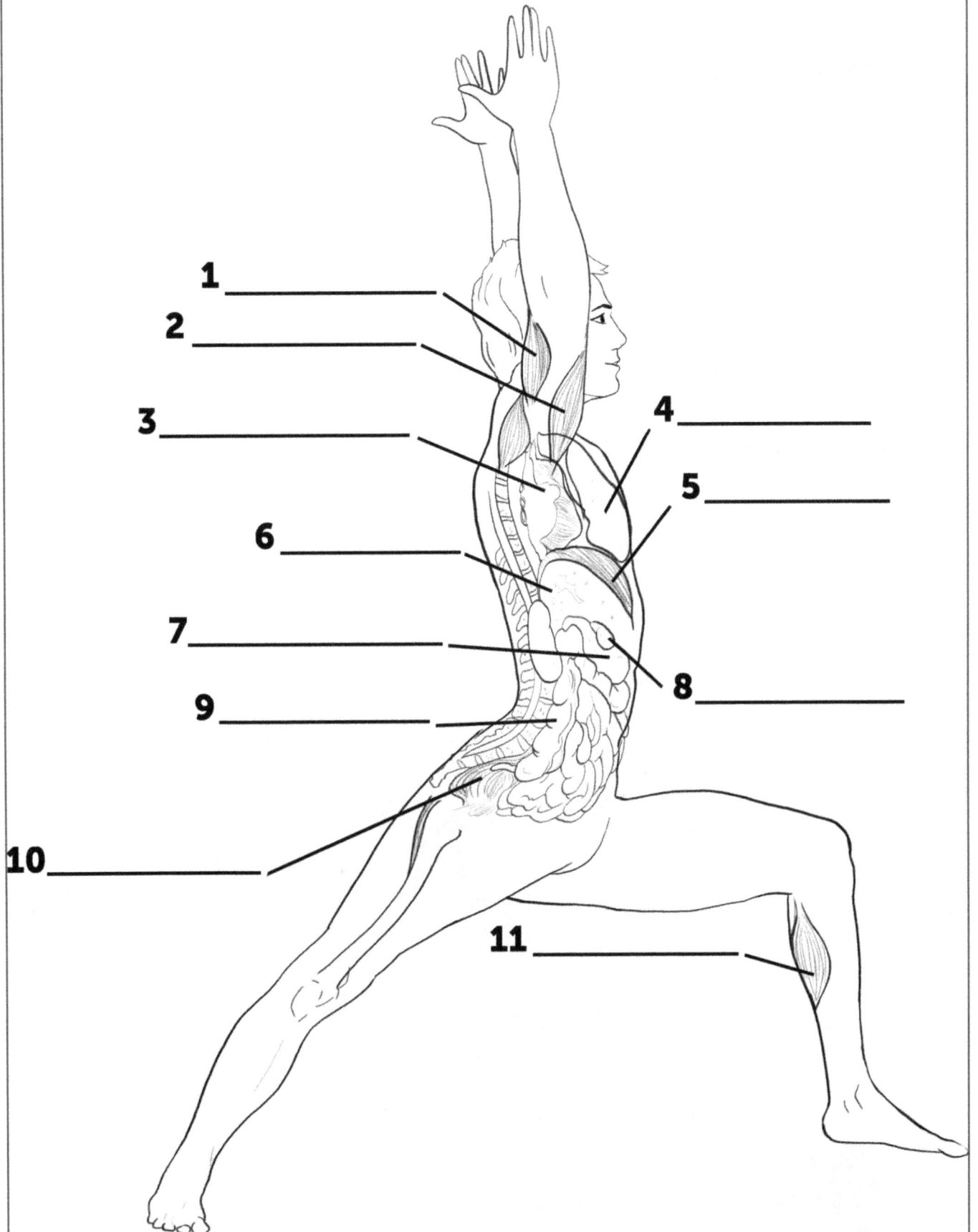

1 _____

2 _____

3 _____

4 _____

5 _____

6 _____

7 _____

8 _____

9 _____

10 _____

11 _____

63. VIRABHADRASANA I

1. BÍCEPS BRAQUIAL
2. TRÍCEPS BRAQUIAL
3. CORAZÓN
4. PULMONES
5. DIAFRAGMA
6. HÍGADO
7. ESTÓMAGO
8. VESÍCULA BILIAR
9. COLON ASCENDENTE
10. RECTO
11. GASTROCNEMIO

64. POSTURA DEL GUERRERO RETORCIDO

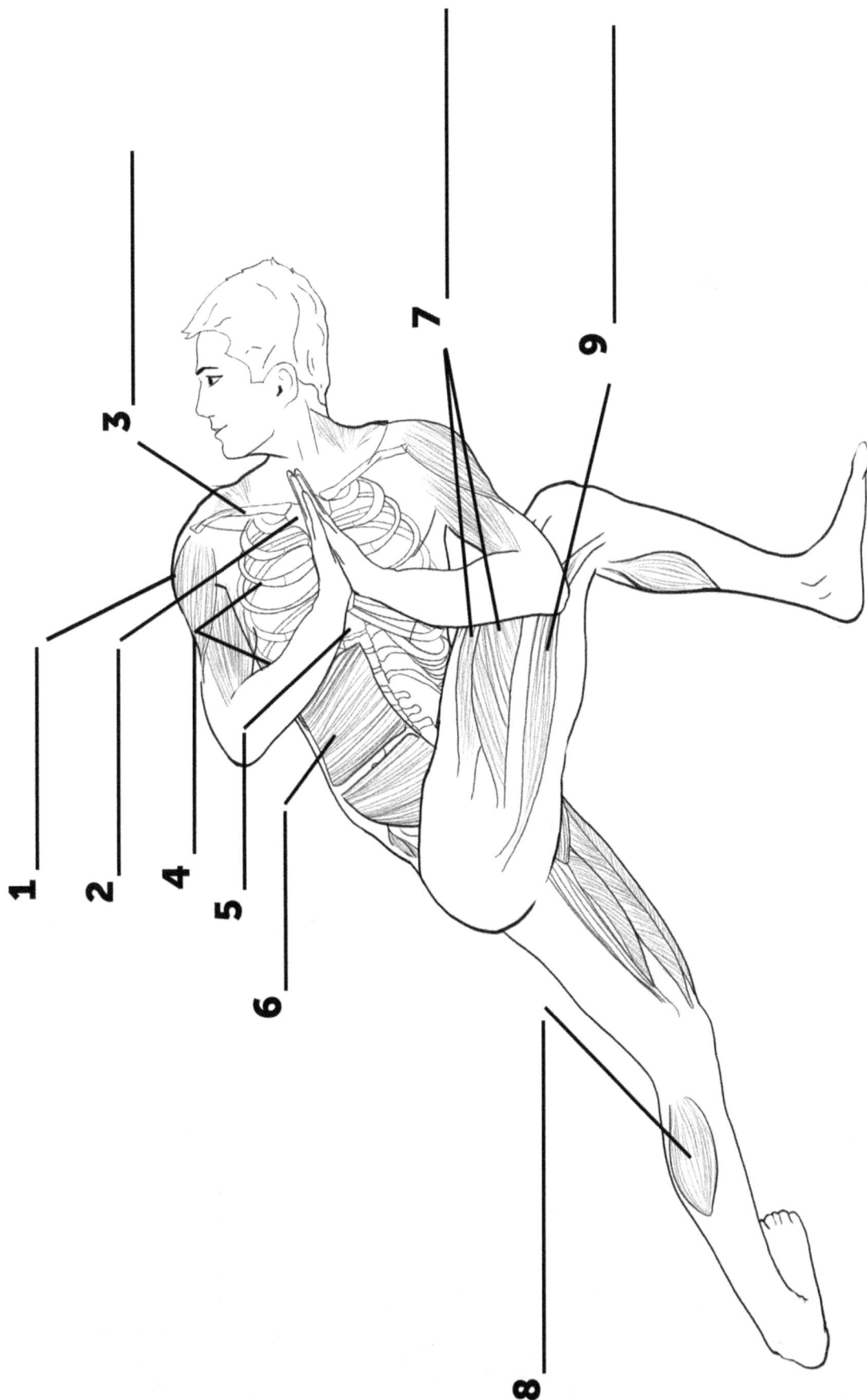

64. POSTURA DEL GUERRERO RETORCIDO

1. DELTOIDES
2. ESTERNÓN
3. CLAVÍCULA
4. COSTILLAS
5. COLUMNA VERTEBRAL
6. OBLICUO INTERNO
7. CUADRÍCEPS
8. GASTROCNEMIO
9. ISQUIOTIBIALES

65. PARIVRTTA TRIKONASANA

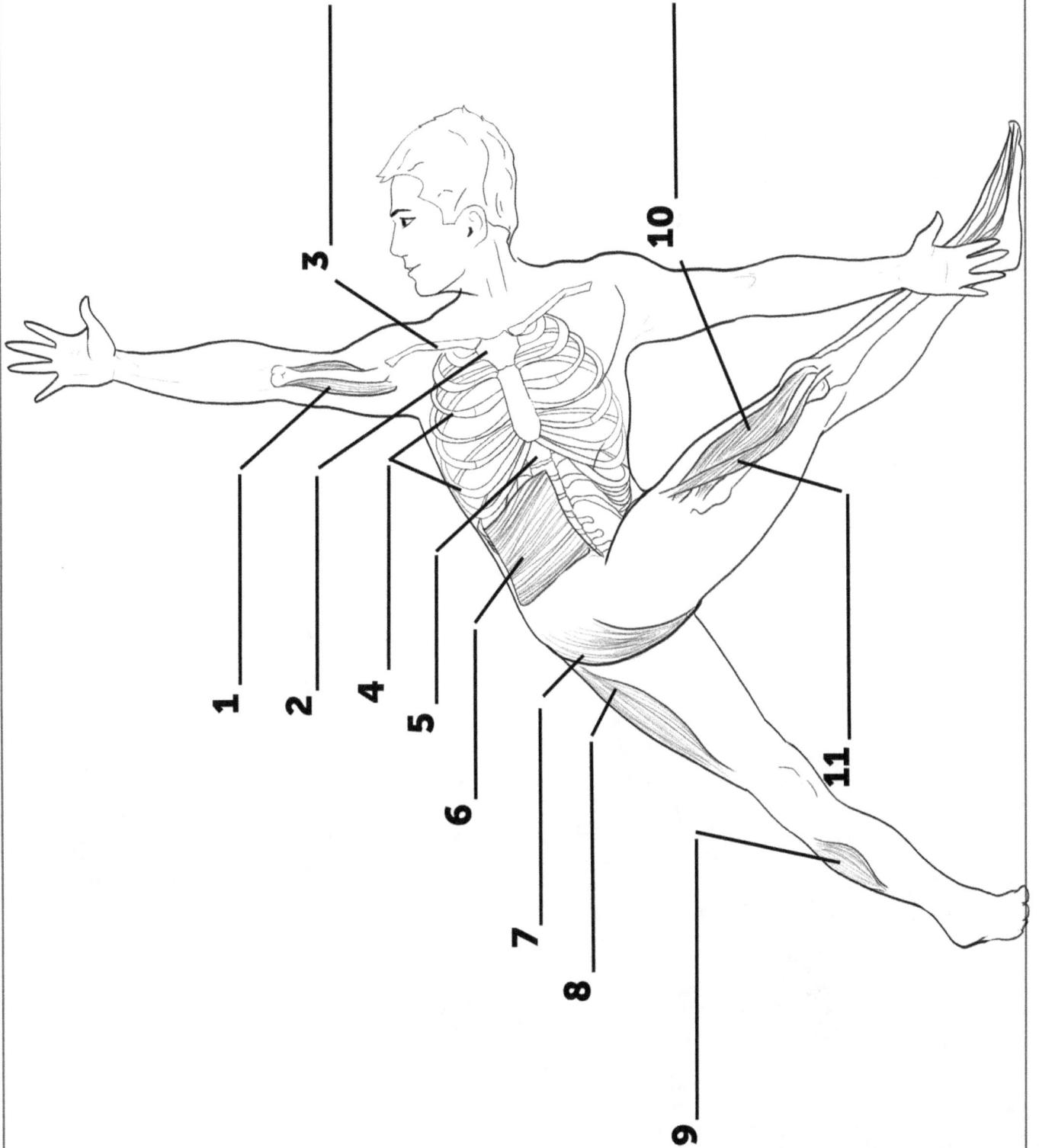

1
2
3
4
5
6
7
8
9
10
11

65. PARIVRTTA TRIKONASANA

1. TRÍCEPS BRAQUIAL
2. ESTERNÓN
3. CLAVÍCULA
4. COSTILLAS
5. COLUMNA VERTEBRAL
6. OBLICUO INTERNO
7. MÚSCULO GLÚTEO MAYOR
8. ISQUIOTIBIALES
9. GASTROCNEMIO
10. CUADRÍCEPS
11. SARTORIO

66. BADDHA PARIVRTTA PARSVAKONASANA

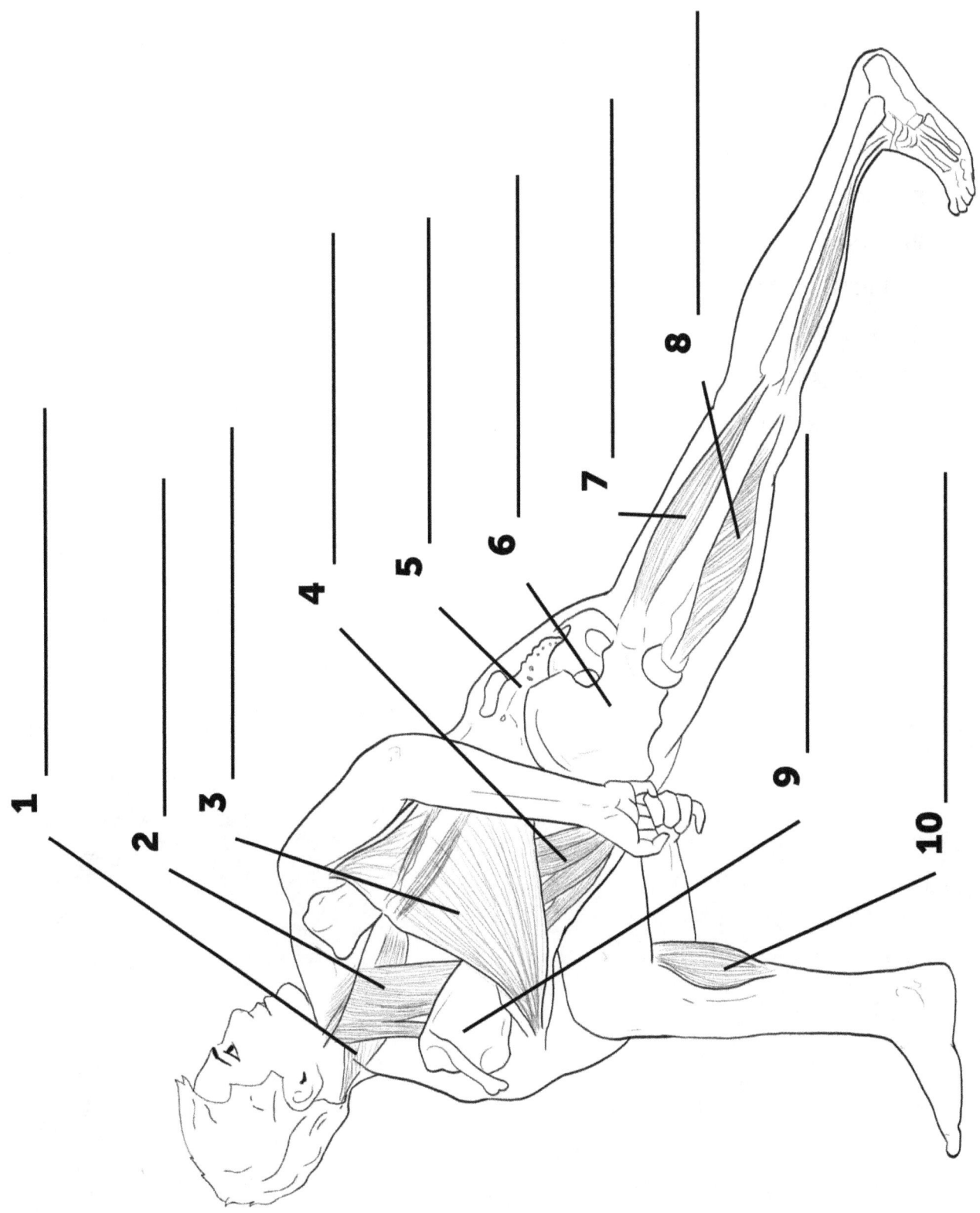

66. BADDHA PARIVRTTA PARSVAKONASANA

1. MÚSCULO ESPLENIO DE LA CABEZA

2. ROMBOIDES

3. LATISSIMUS DORSI

4. ERECTOR DE LA COLUMNA

5. SACRO

6. PELVIS

7. ISQUIOTIBIALES

8. CUADRÍCEPS

9. ESCÁPULA

10. GASTROCNEMIO

67. POSTURA DEL CAMELLO

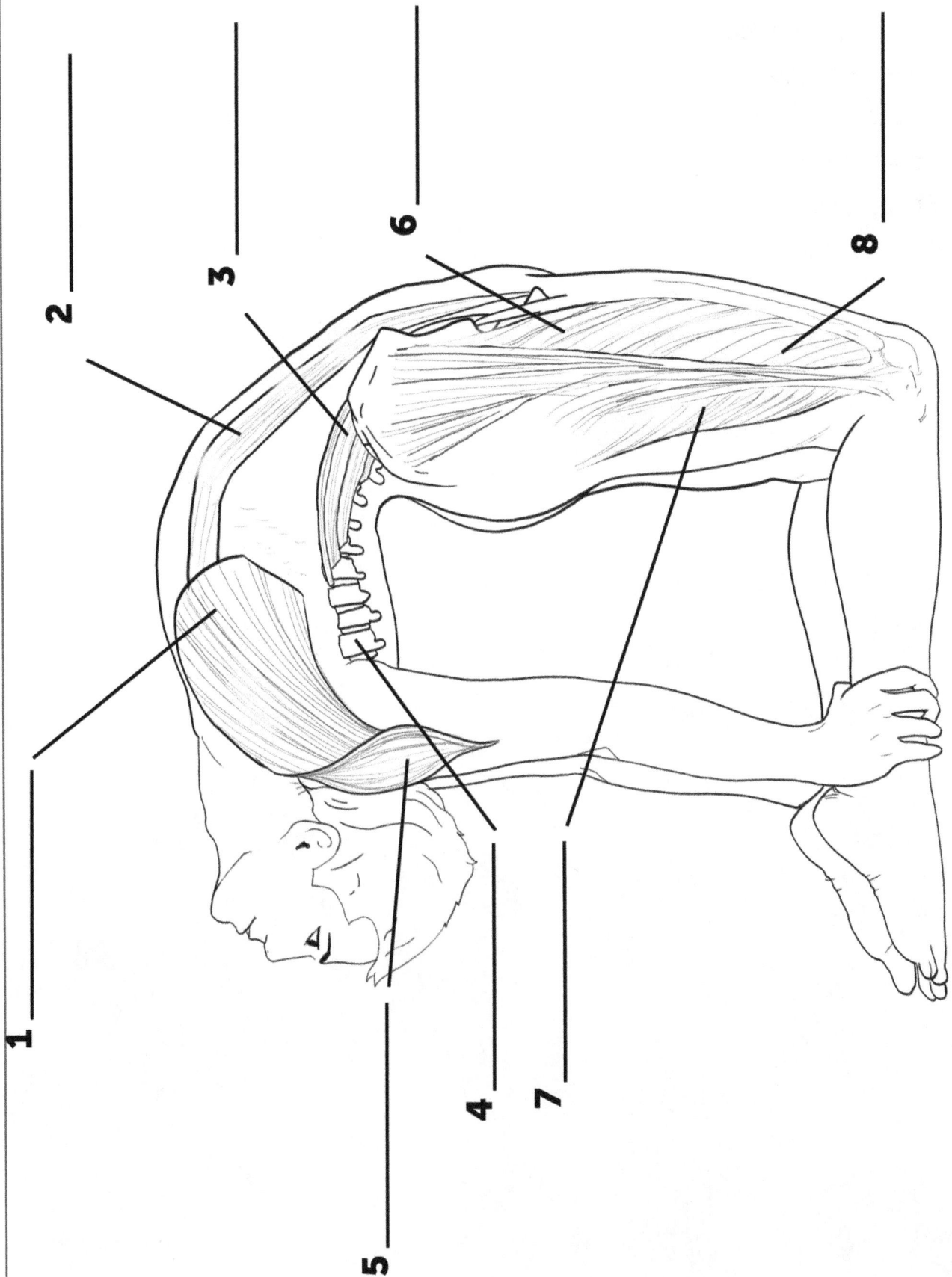

67. POSTURA DEL CAMELLO

1.   PECTORAL MAYOR

2.   RECTO ABDOMINAL

3.   MÚSCULO PSOAS MAYOR

4.   COLUMNA VERTEBRAL

5.   DELTOIDES

6.   RECTO FEMORAL

7.   ISQUIOTIBIALES

8.   MÚSCULO VASTO LATERAL

68. VIRABHADRASANA I

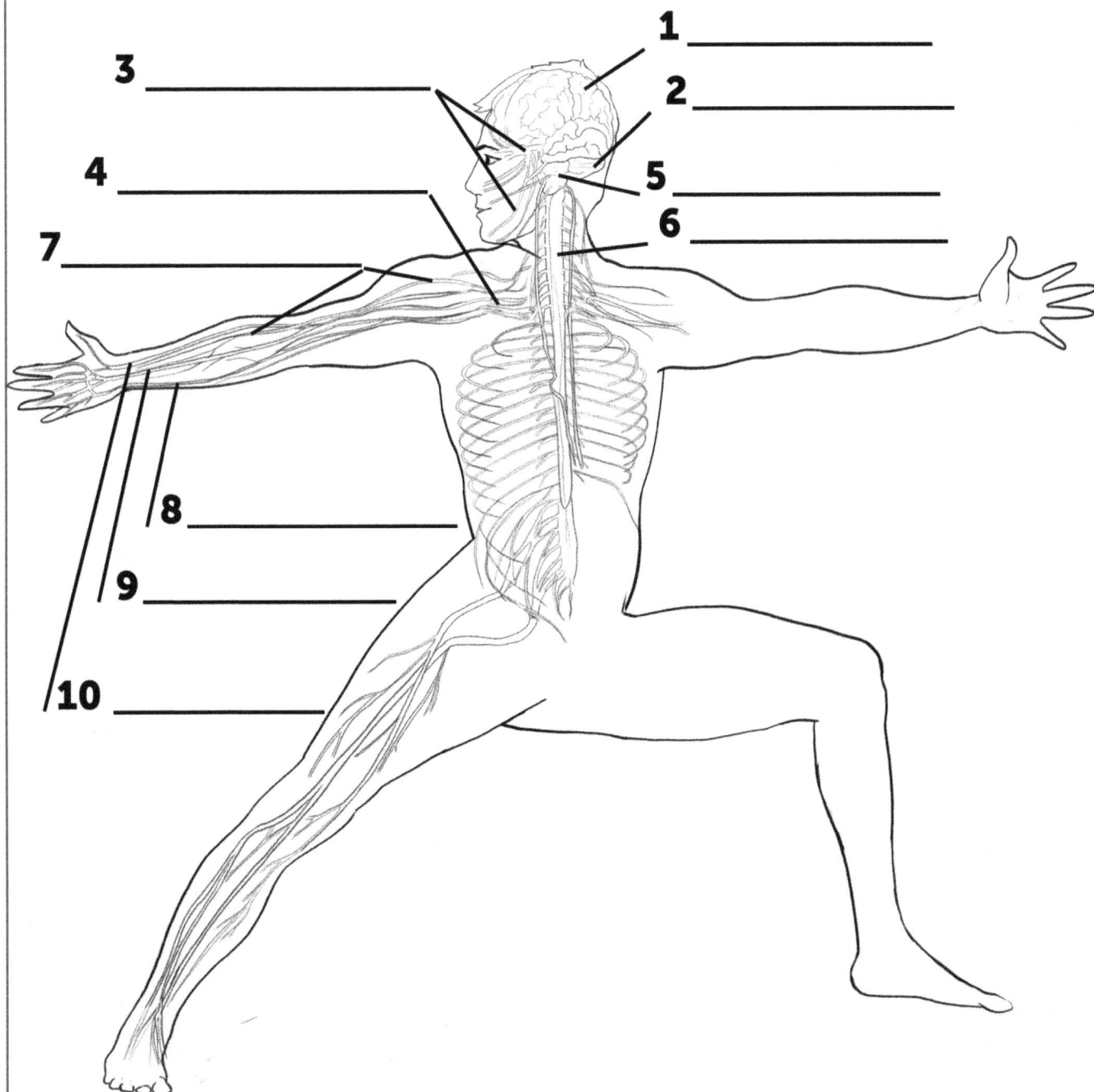

1 _____

2 _____

3 _____

4 _____

5 _____

6 _____

7 _____

8 _____

9 _____

10 _____

68. VIRABHADRASANA I

1. CEREBRO
2. CEREBELO
3. NERVIOS CRANEALES
4. PLEXO BRAQUIAL
5. TRONCO ENCEFÁLICO
6. MÉDULA ESPINAL
7. MUSCULOCUTÁNEO
8. ULNAR
9. MEDIANA
10. RADIAL

69. VIRABHADRASANA I

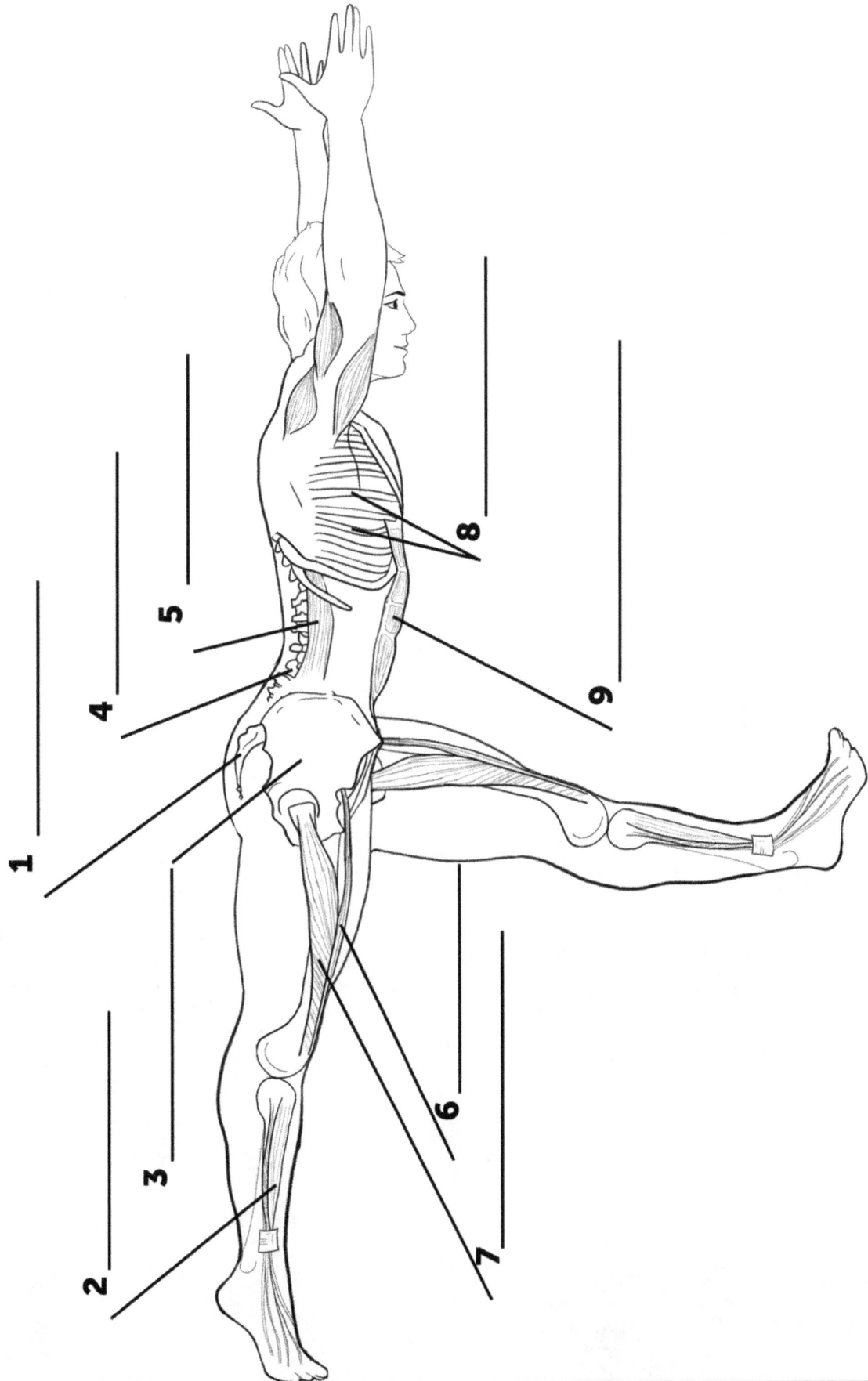

69. VIRABHADRASANA I

1. SACRO
2. TIBIAL ANTERIOR
3. PELVIS
4. COLUMNA VERTEBRAL
5. ERECTOR DE LA COLUMNA
6. SARTORIO
7. RECTO FEMORAL
8. COSTILLAS
9. RECTO ABDOMINAL

70. VIPARITA VIRABHADRASANA

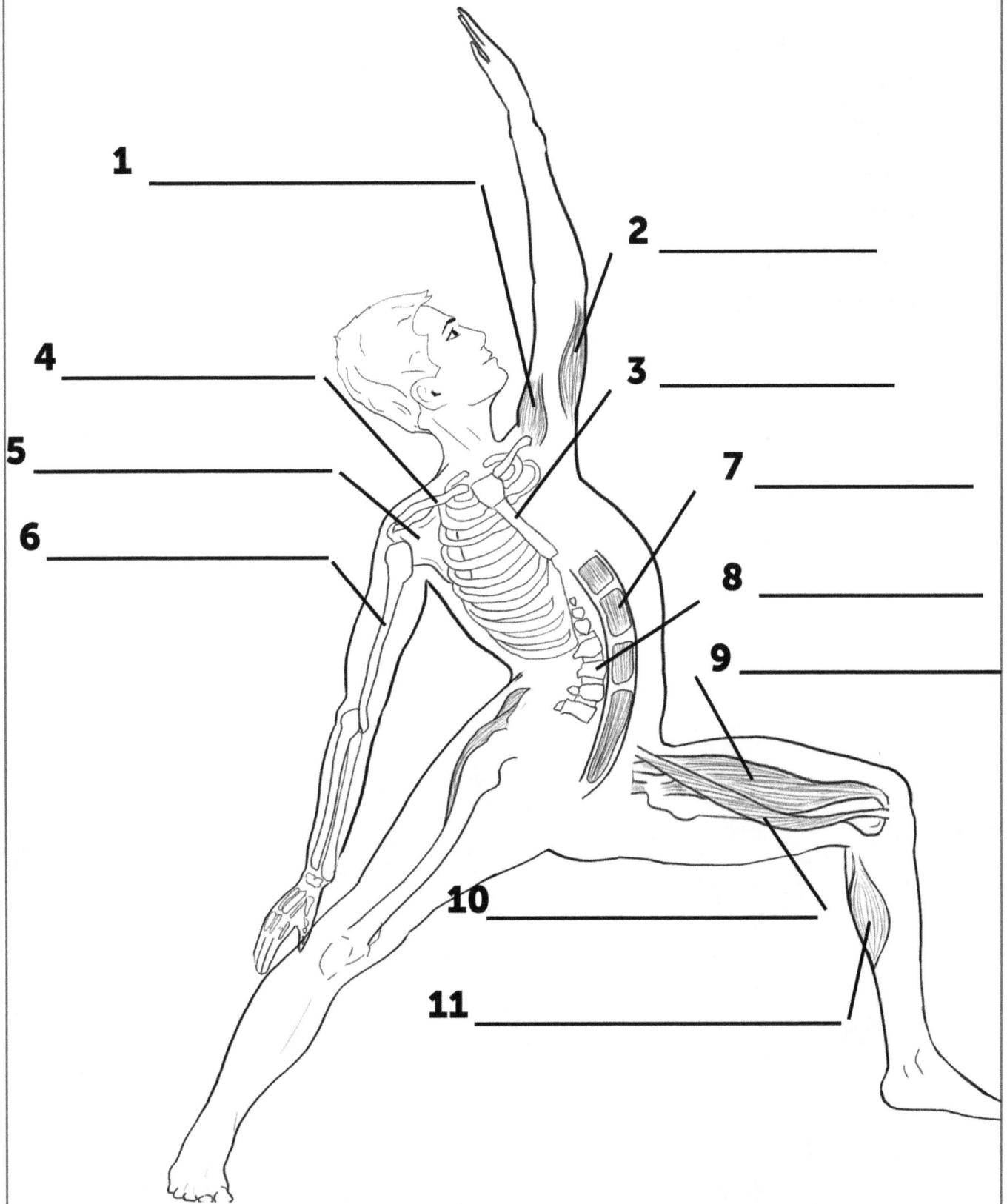

1 _____

2 _____

4 _____

3 _____

5 _____

7 _____

6 _____

8 _____

9 _____

10 _____

11 _____

70. VIPARITA VIRABHADRASANA

1. DELTOIDES
2. TRÍCEPS BRAQUIAL
3. ESTERNÓN
4. CLAVÍCULA
5. ESCÁPULA
6. HÚMERO
7. RECTO ABDOMINAL
8. COLUMNA VERTEBRAL
9. RECTO FEMORAL
10. SARTORIO
11. GASTROCNEMIO

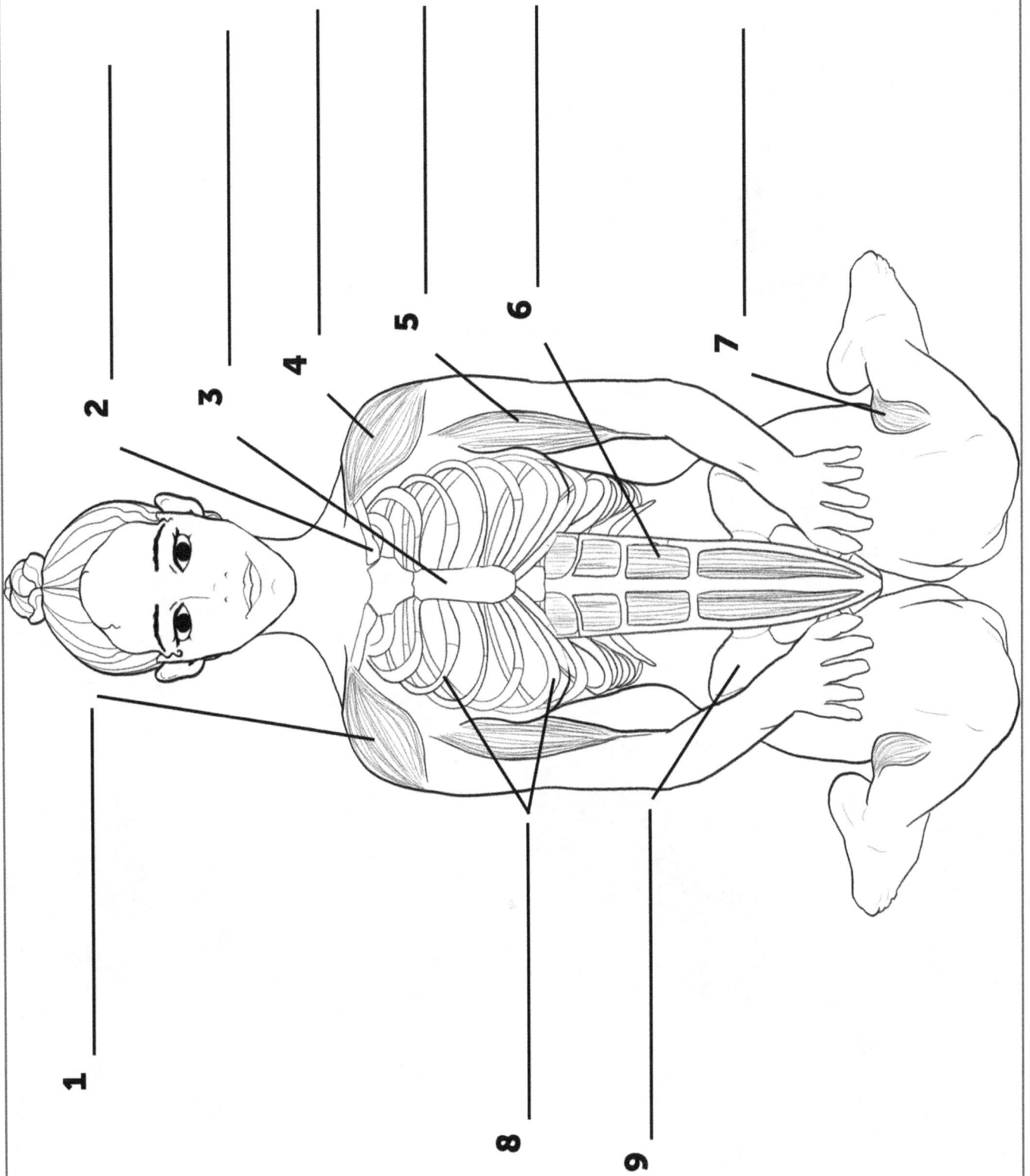

71. VIRASANA

1

2

3

4

5

6

7

8

9

71. VIRASANA

1.   DELTOIDES
2.   CLAVÍCULA
3.   ESTERNÓN
4.   DELTOIDES
5.   BÍCEPS BRAQUIAL
6.   RECTO ABDOMINAL
7.   GASTROCNEMIO
8.   COSTILLAS
9.   PELVIS

72. ARDHA SUPTA VIRASANA

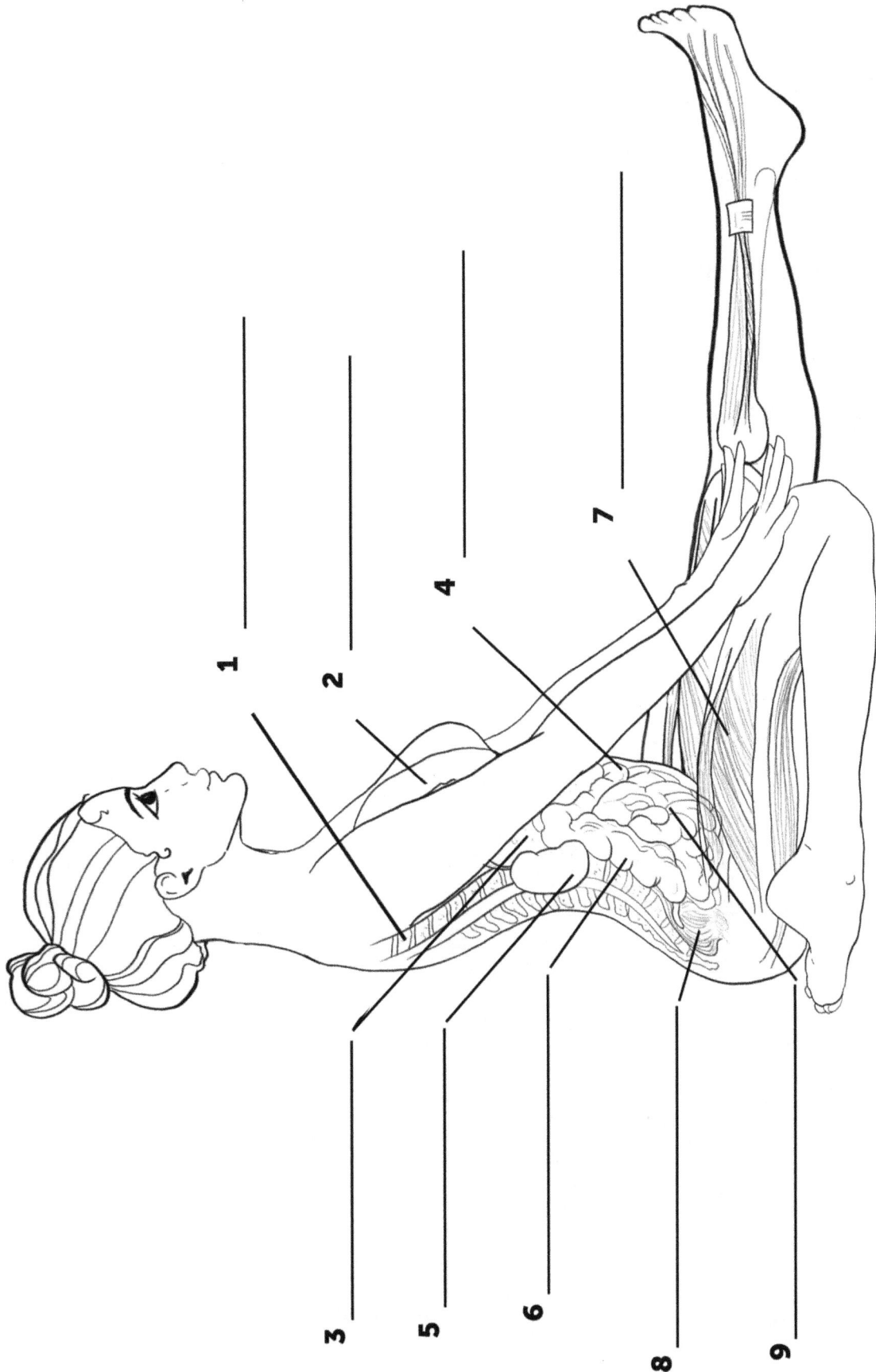

1

2

3

4

5

6

7

8

9

72. ARDHA SUPTA VIRASANA

1. COLUMNA VERTEBRAL

2. PULMONES

3. HÍGADO

4. COLON TRANSVERSO

5. RIÑÓN

6. COLON ASCENDENTE

7. CUADRÍCEPS

8. RECTO

9. FOLICULOS DE INTESTINO DELGADO

73. SUPTA VIRASANA

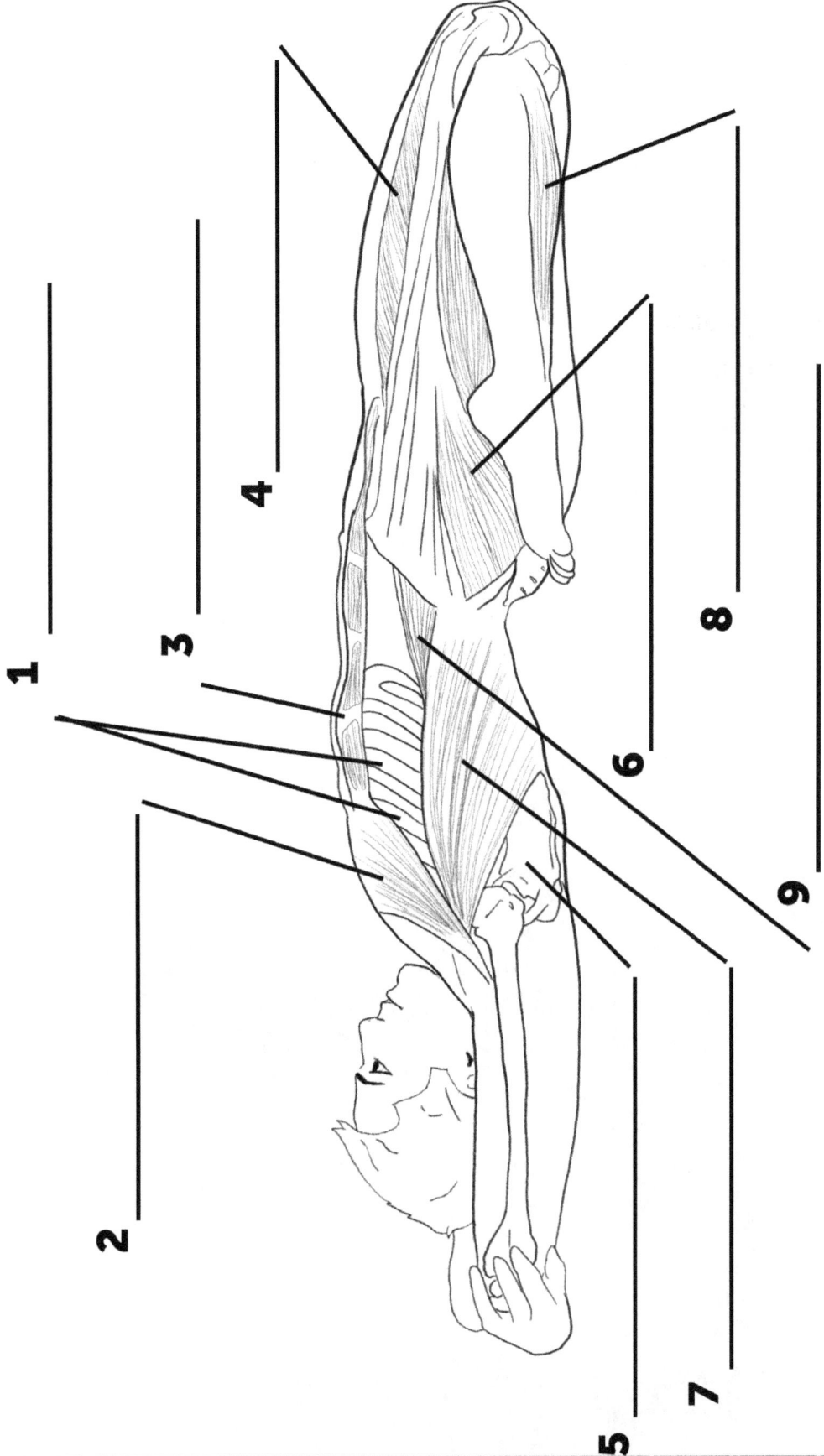

73. SUPTA VIRASANA

1.   COSTILLAS
2.   PECTORAL MAYOR
3.   RECTO ABDOMINAL
4.   MÚSCULO VASTO LATERAL
5.   ESCÁPULA
6.   MÚSCULO GLÚTEO MAYOR
7.   LATISSIMUS DORSI
8.   TIBIAL ANTERIOR
9.   MÚSCULO PSOAS MAYOR

74. UTTHITA HASTA PADANGUSTHASANA

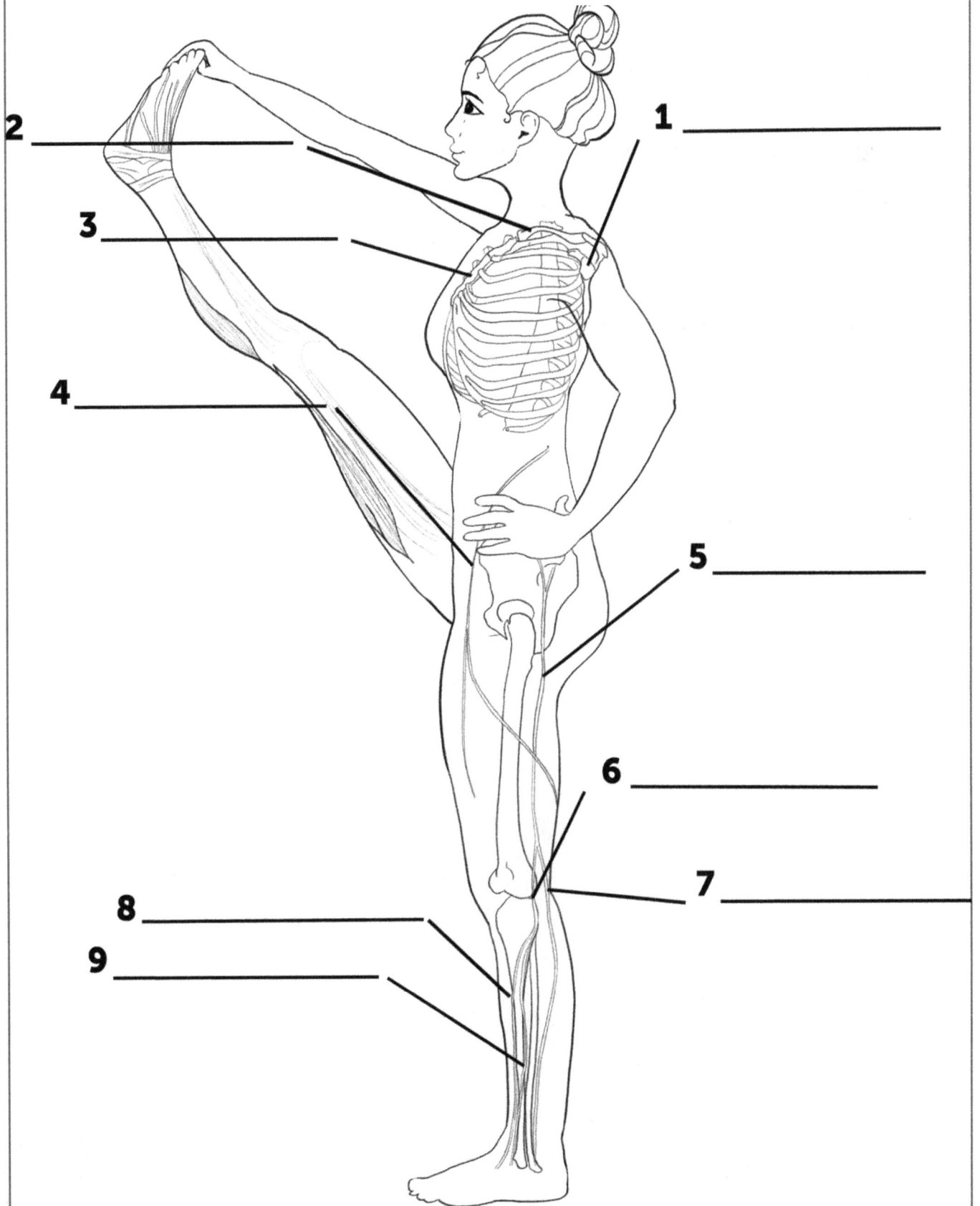

1 _____

2 _____

3 _____

4 _____

5 _____

6 _____

7 _____

8 _____

9 _____

74. UTTHITA HASTA PADANGUSTHASANA

1. ESCÁPULA
2. CLAVÍCULA
3. ESTERNÓN
4. NERVIO CUTÁNEO FEMORAL LATERAL
5. NERVIO CIÁTICO
6. NERVIO PERONEO COMÚN
7. NERVIO TIBIAL
8. NERVIO PERONEO PROFUNDO
9. NERVIO PERONEO SUPERFICIAL

75. KAPOTASANA

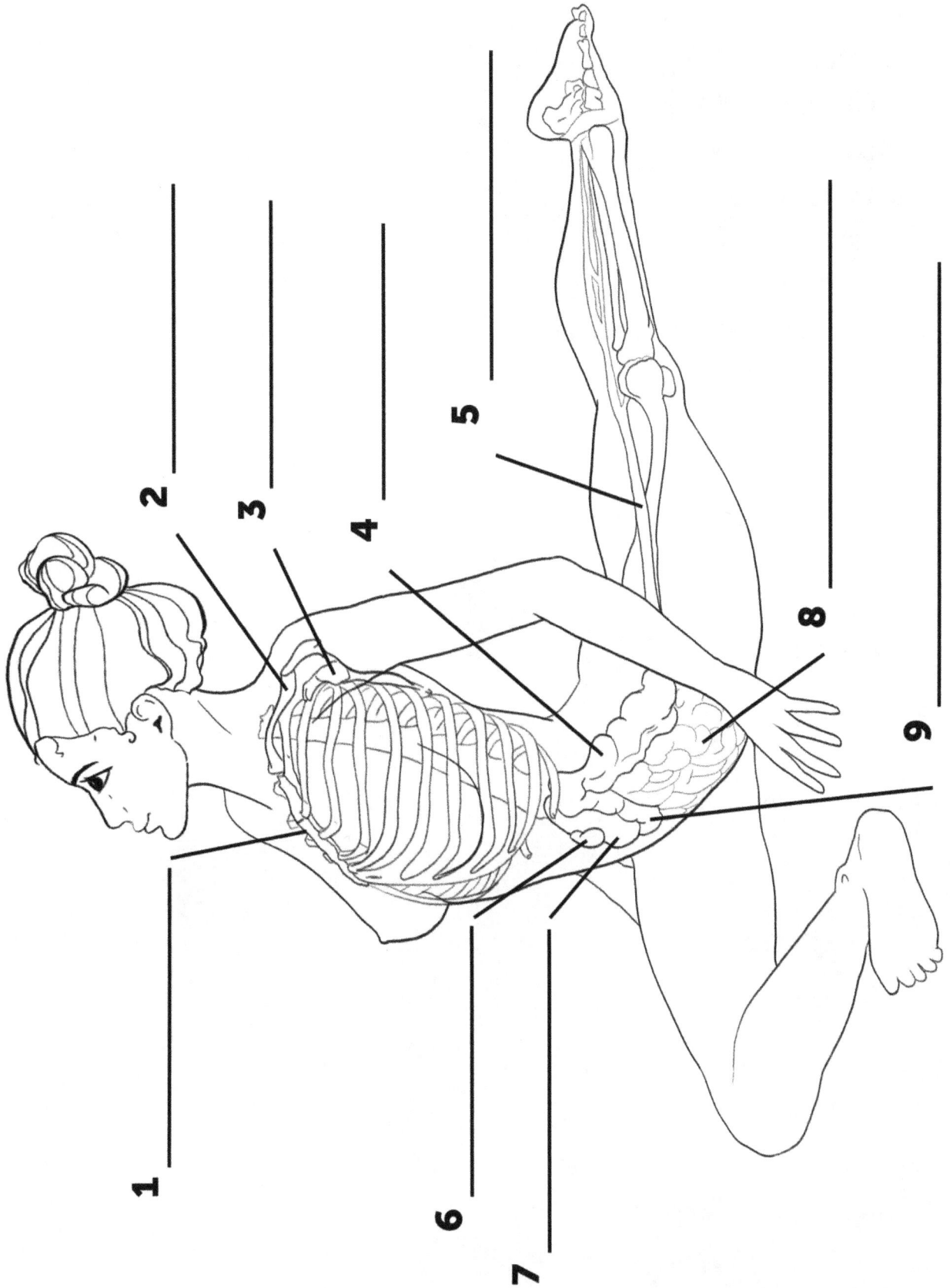

1

2

3

4

5

6

7

8

9

75. KAPOTASANA

1. ESTERNÓN
2. CLAVÍCULA
3. ESCÁPULA
4. COLON ASCENDENTE
5. NERVIO CIÁTICO
6. VESÍCULA BILIAR
7. ESTÓMAGO
8. FOLICULOS DE INTESTINO DELGADO
9. COLON TRANSVERSO

76. URDHVA MUKHA PASASANA VINYASA

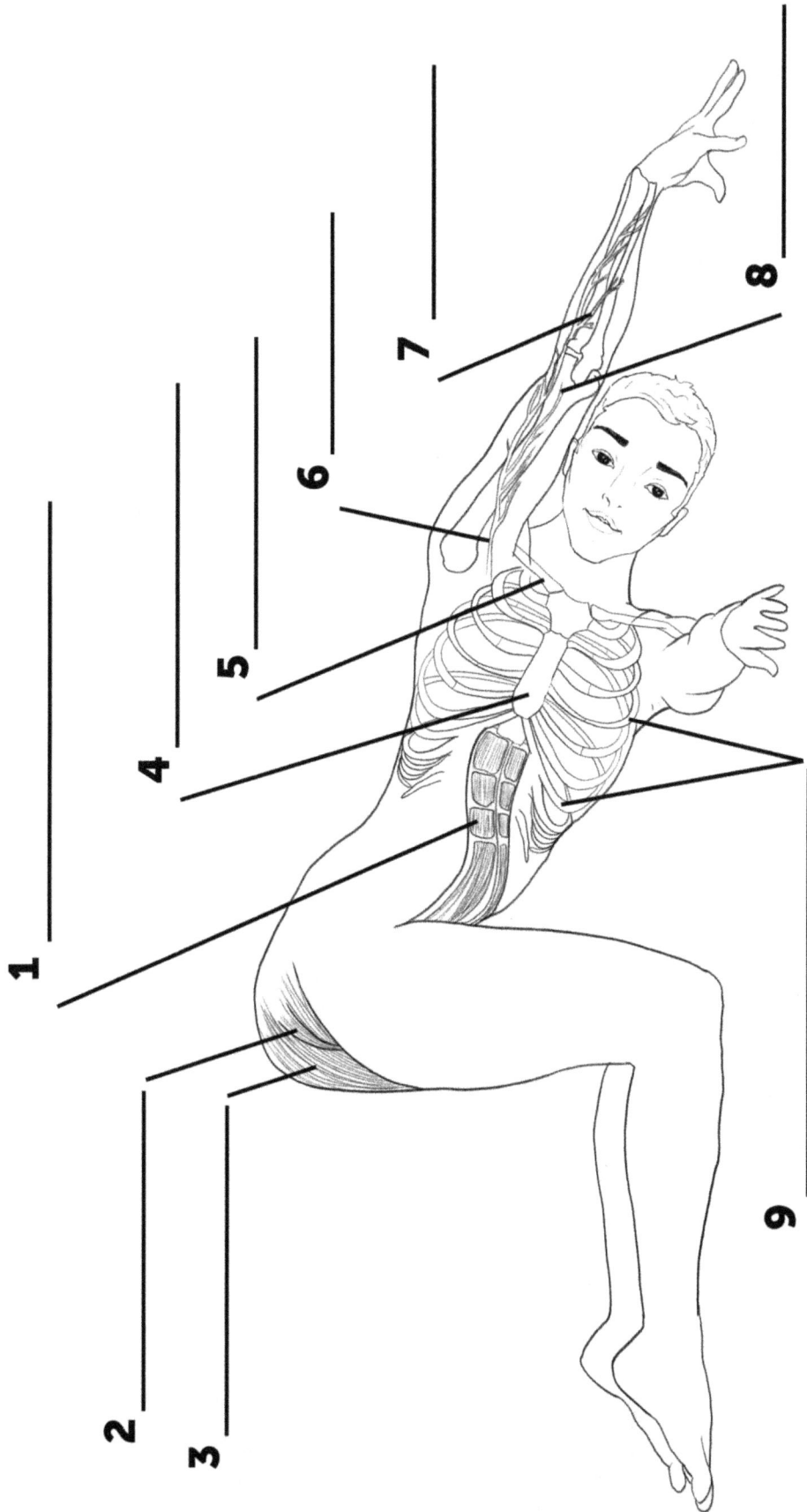

1

2

3

4

5

6

7

8

9

76. URDHVA MUKHA PASASANA VINYASA

1. RECTO ABDOMINAL
2. PIRIFORME
3. MÚSCULO GLÚTEO MAYOR
4. ESTERNÓN
5. CLAVÍCULA
6. NERVIO RADIAL
7. NERVIO INTERÓSEO POSTERIOR
8. ANCÓNEO
9. COSTILLAS

77. KROUNCHASANA

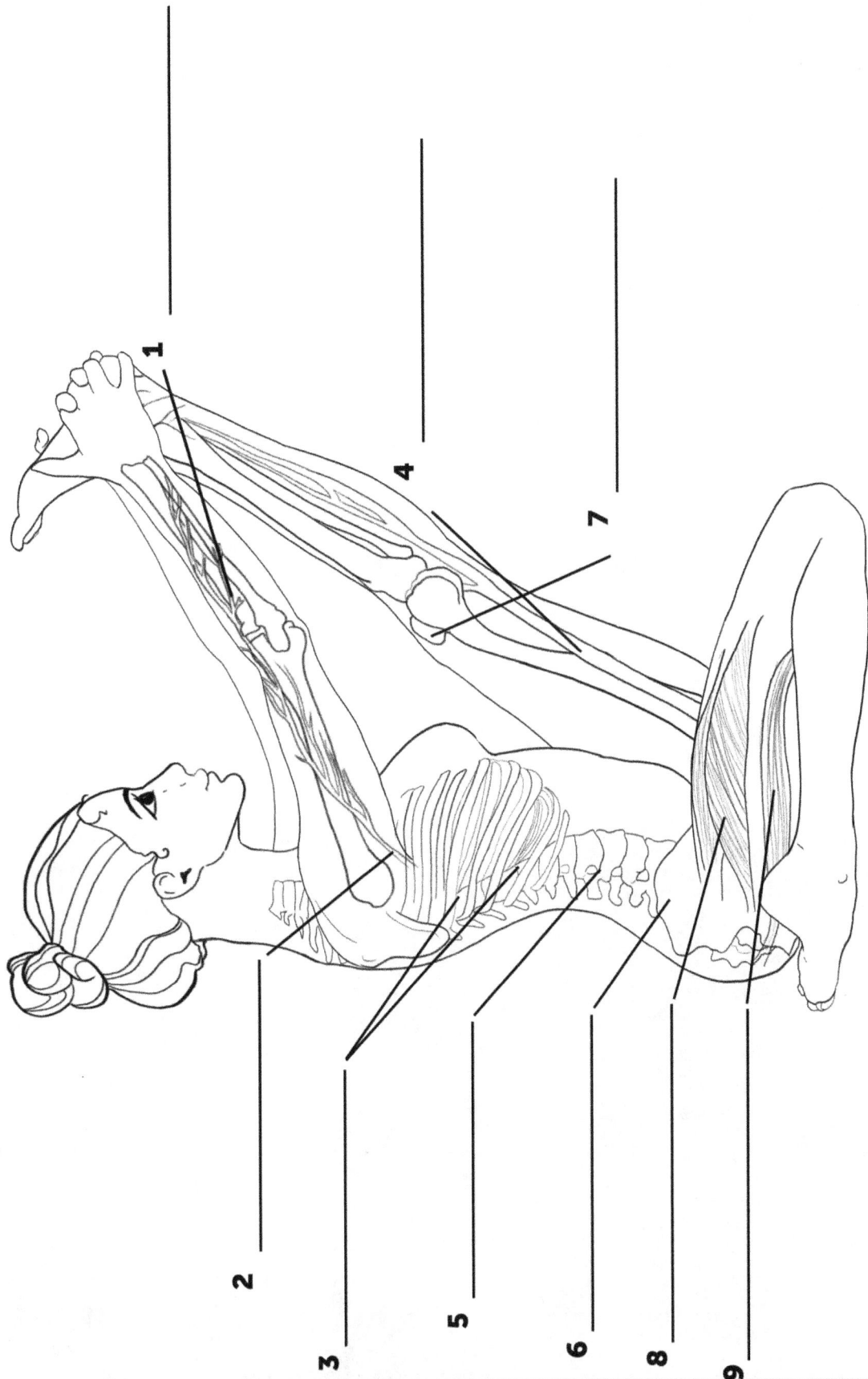

77. KROUNCHASANA

1. NERVIO INTERÓSEO POSTERIOR
2. NERVIO RADIAL
3. COSTILLAS
4. NERVIO CIÁTICO
5. COLUMNA VERTEBRAL
6. PELVIS
7. RÓTULA
8. CUADRÍCEPS
9. ISQUIOTIBIALES

78. DHANURASANA

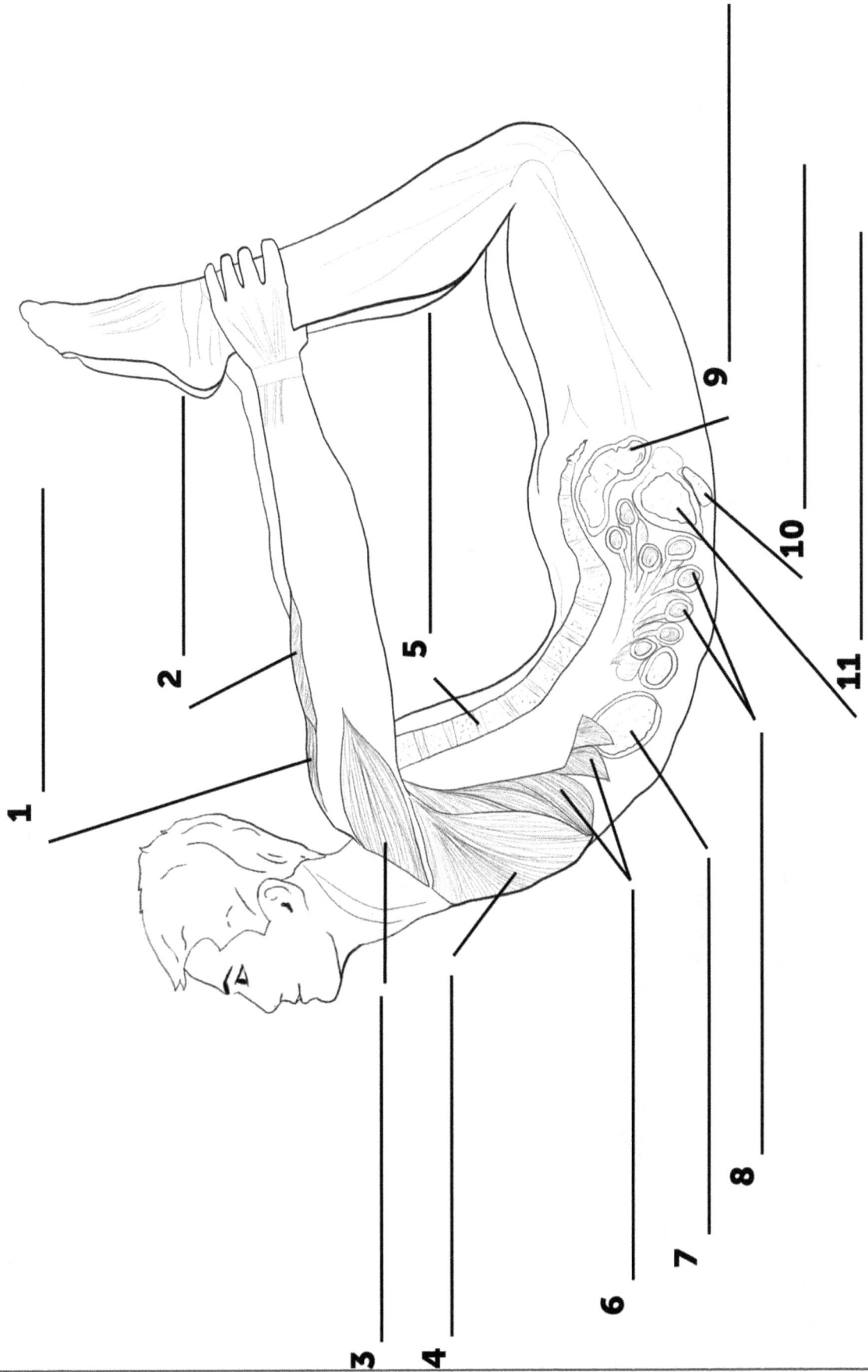

1
2
3
4
5
6
7
8
9
10
11

78. DHANURASANA

1. DELTOIDES POSTERIOR
2. TRÍCEPS BRAQUIAL
3. DELTOIDES ANTERIOR
4. PECTORAL MAYOR
5. COLUMNA VERTEBRAL
6. SERRATO ANTERIOR
7. ESTÓMAGO
8. FOLICULOS DE INTESTINO DELGADO
9. RECTO
10. HUESO PÚBICO
11. VEJIGA URINARIA

79. URDHVA DHANURASANA

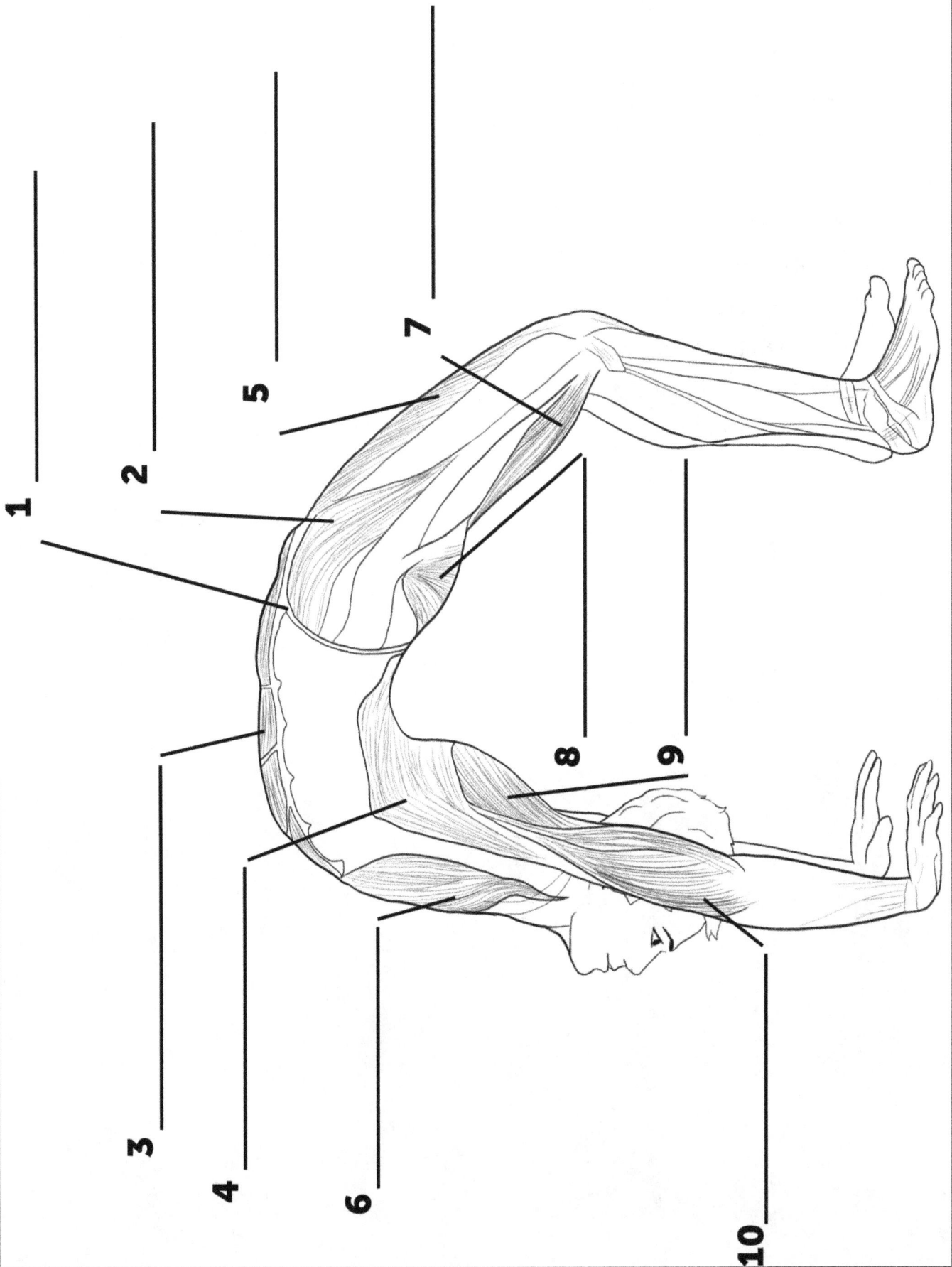

1

2

3

4

5

6

7

8

9

10

79. URDHVA DHANURASANA

1. ILIOPSOAS
2. MÚSCULO TENSOR DE LA FASCIA LATA
3. RECTO ABDOMINAL
4. LATISSIMUS DORSI
5. CUADRÍCEPS
6. PECTORAL MAYOR
7. ISQUIOTIBIALES
8. MÚSCULO GLÚTEO MAYOR
9. ERECTOR DE LA COLUMNA
10. TRÍCEPS BRAQUIAL

80. UTTHAN PRISTHASANA

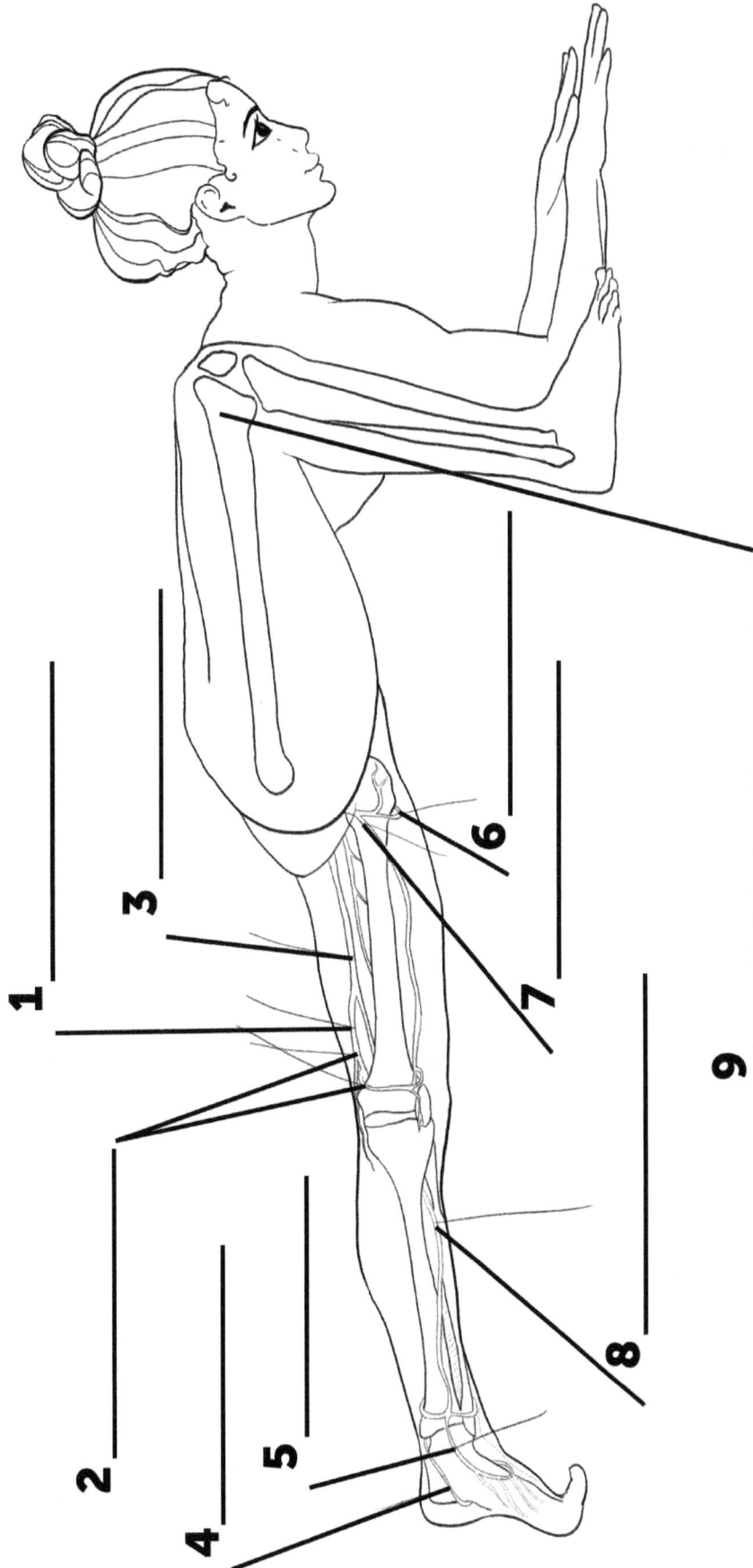

1

2

3

4

5

6

7

8

9

80. UTTHAN PRISTHASANA

1. HIATO ADUCTOR
2. ARTERIAS GENICULARES
3. ARTERIA FEMORAL
4. ARTERIA PLANTAR MEDIAL
5. ARTERIA DORSALIS PEDIS
6. ARTERIA FEMORAL CIRCUNFLEJA LATERAL
7. RAMA DESCENDENTE
8. ARTERIA TIBIAL ANTERIOR
9. FÉMUR

81. EKA PADA RAJAKAPOTASANA

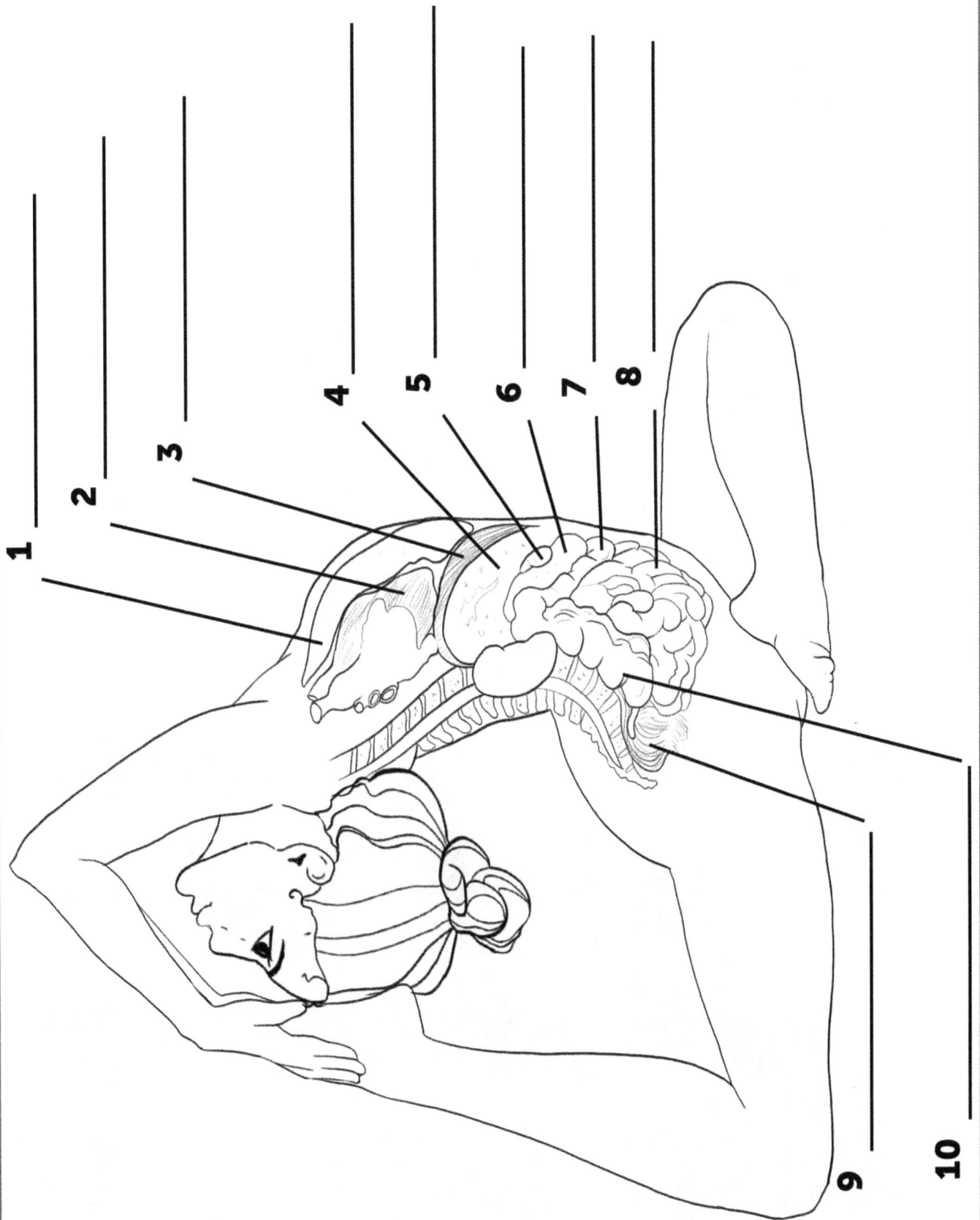

1

2

3

4

5

6

7

8

9

10

81. EKA PADA RAJAKAPOTASANA

1. PULMONES
2. CORAZÓN
3. DIAFRAGMA
4. HÍGADO
5. VESÍCULA BILIAR
6. ESTÓMAGO
7. COLON TRANSVERSO
8. FOLICULOS DE INTESTINO DELGADO
9. RECTO
10. COLON ASCENDENTE

82. VRIKSASANA

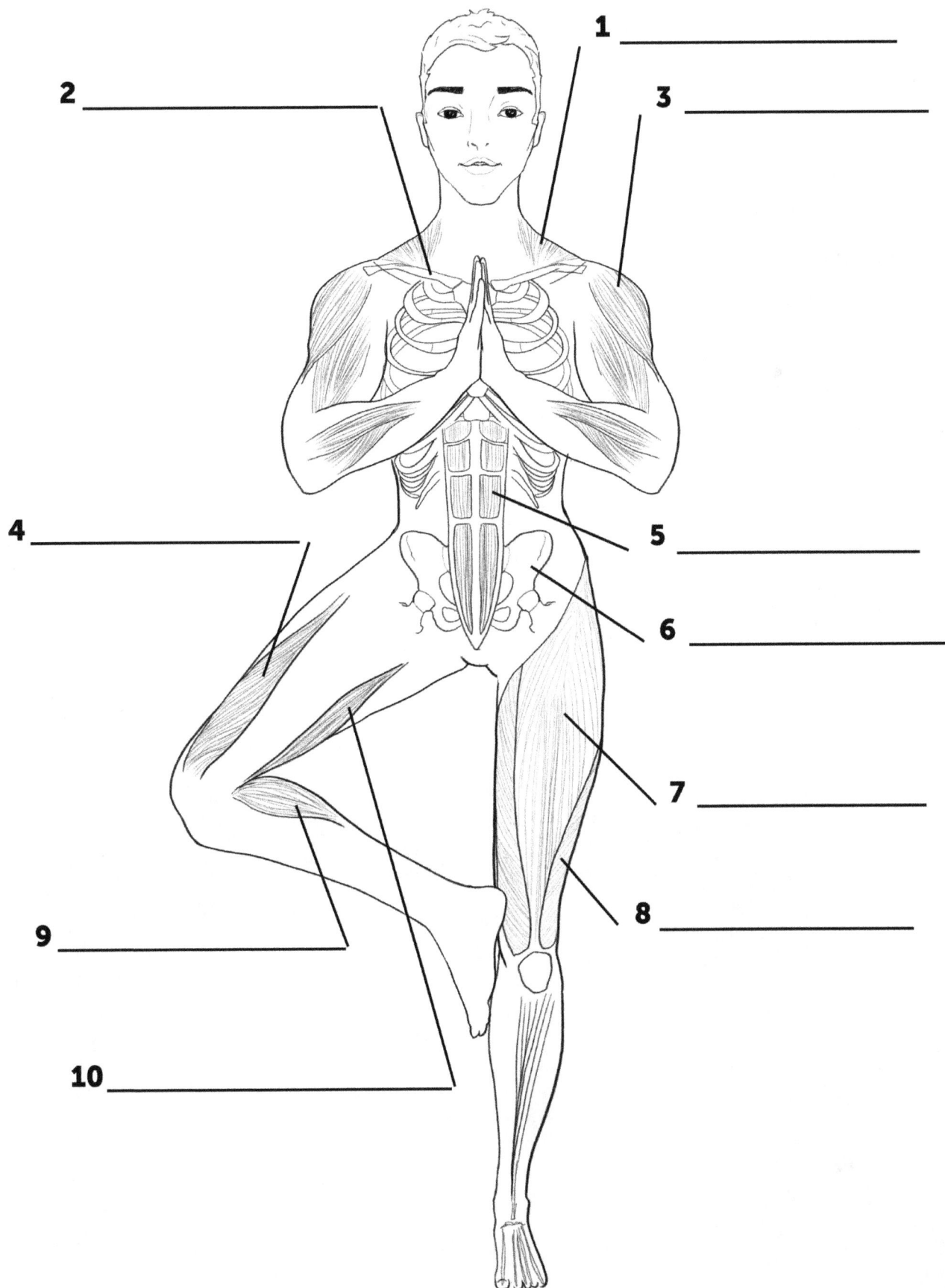

1 _____

2 _____

3 _____

4 _____

5 _____

6 _____

7 _____

8 _____

9 _____

10 _____

82. VRIKSASANA

1. TRAPECIO
2. CLAVÍCULA
3. DELTOIDES
4. CUADRÍCEPS
5. RECTO ABDOMINAL
6. PELVIS
7. RECTO FEMORAL
8. MÚSCULO VASTO LATERAL
9. GASTROCNEMIO
10. ISQUIOTIBIALES

83. POSTURA DEL CUERVO

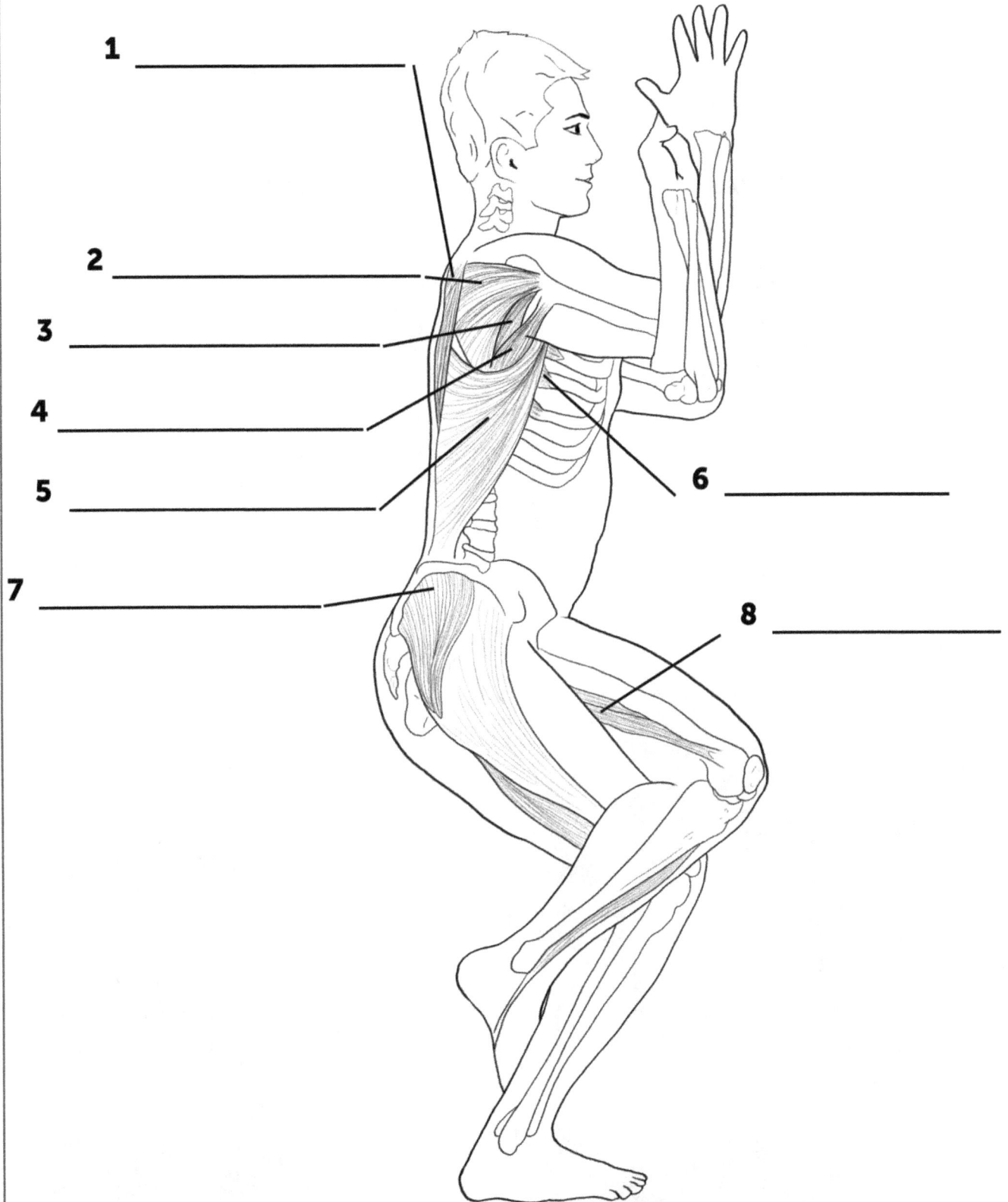

1 _____

2 _____

3 _____

4 _____

5 _____

6 _____

7 _____

8 _____

83. POSTURA DEL CUERVO

1. TRAPECIO
2. INFRAESPINOSO
3. REDONDO MENOR
4. MÚSCULO REDONDO MAYOR
5. LATISSIMUS DORSI
6. SERRATO ANTERIOR
7. GLÚTEO MEDIO
8. ADUCTOR MAYOR

84. JANU SIRSASANA

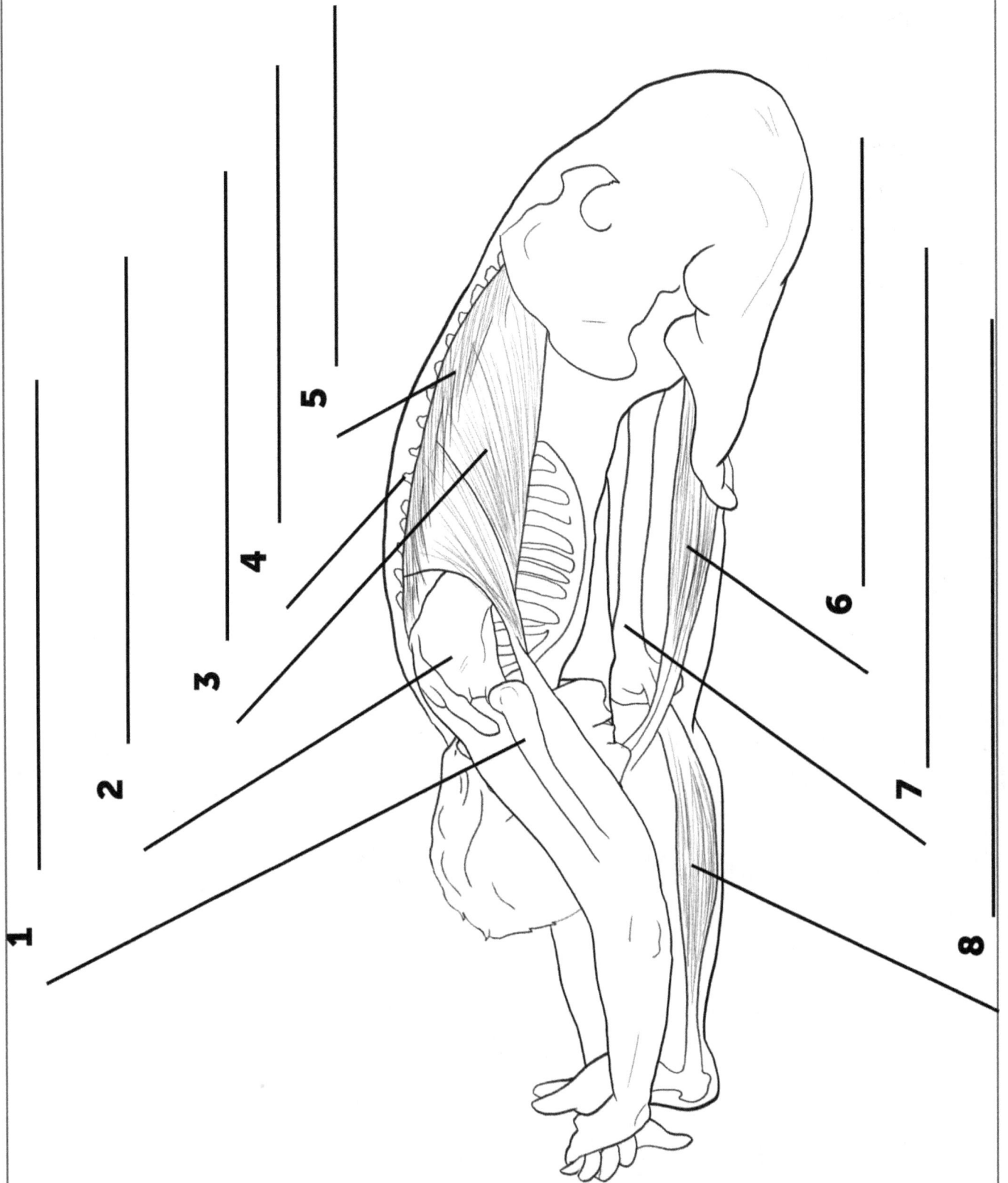

84. JANU SIRSASANA

1. HÚMERO
2. ESCÁPULA
3. LATISSIMUS DORSI
4. COLUMNA VERTEBRAL
5. ERECTOR DE LA COLUMNA
6. ISQUIOTIBIALES
7. FÉMUR
8. GASTROCNEMIO

85. POSTURA DEL BAILARIN

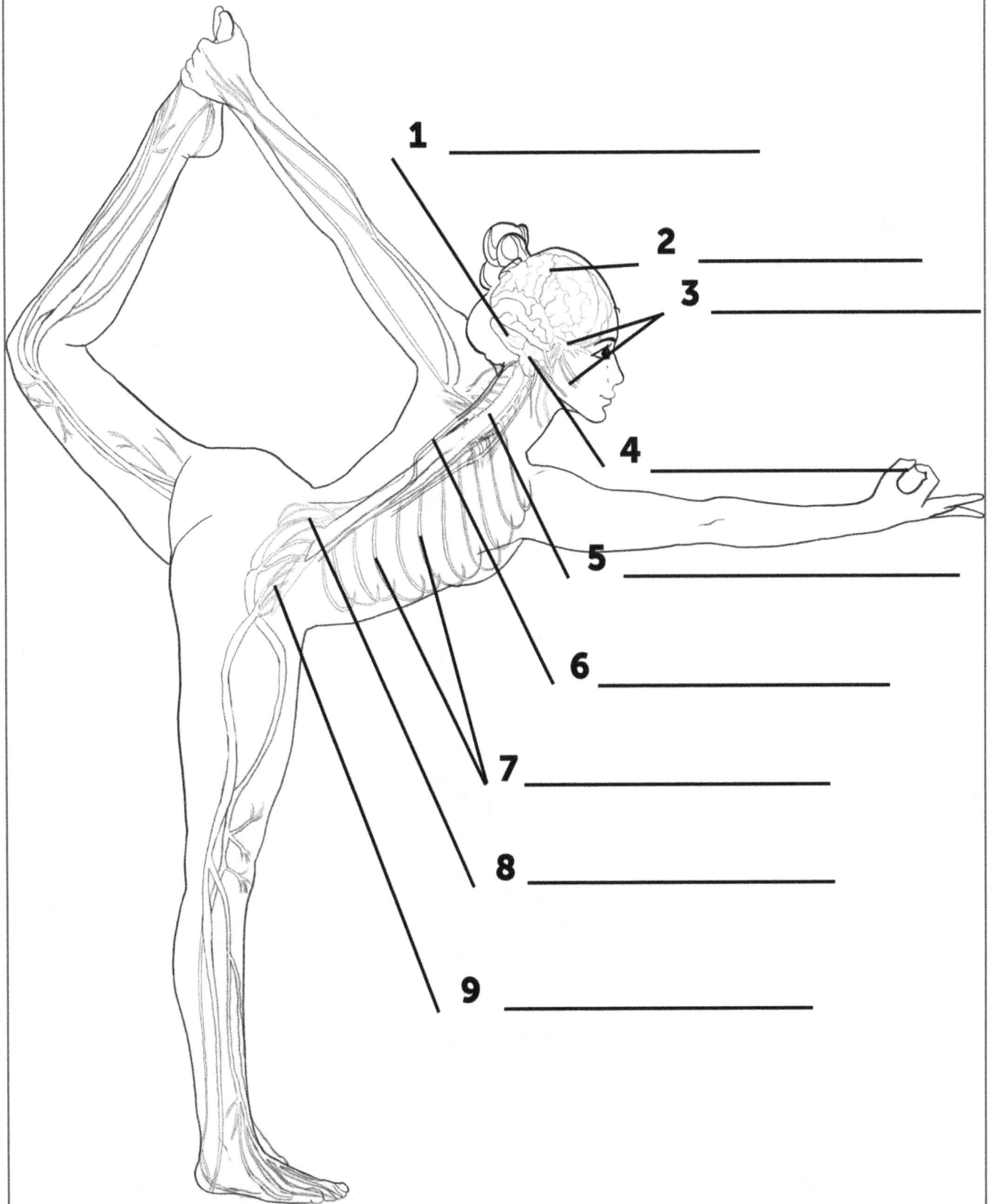

1 _____

2 _____

3 _____

4 _____

5 _____

6 _____

7 _____

8 _____

9 _____

85. POSTURA DEL BAILARIN

1. CEREBELO
2. CEREBRO
3. NERVIOS CRANEALES
4. TRONCO ENCEFÁLICO
5. MÉDULA ESPINAL
6. NERVIO VAGO
7. INTERCOSTALES
8. PLEXO LUMBAR
9. PLEXO SACRO

86. PARIVRTTA UTKATASANA

1 _____

2 _____

3 _____

4 _____

5 _____

6 _____

7 _____

8 _____

9 _____

86. PARIVRTTA UTKATASANA

1. AORTA
2. CORAZÓN
3. PULMONES
4. HÍGADO
5. ESTÓMAGO
6. COLON ASCENDENTE
7. FOLICULOS DE INTESTINO DELGADO
8. ISQUIOTIBIALES
9. GASTROCNEMIO

87. SASANGASANA

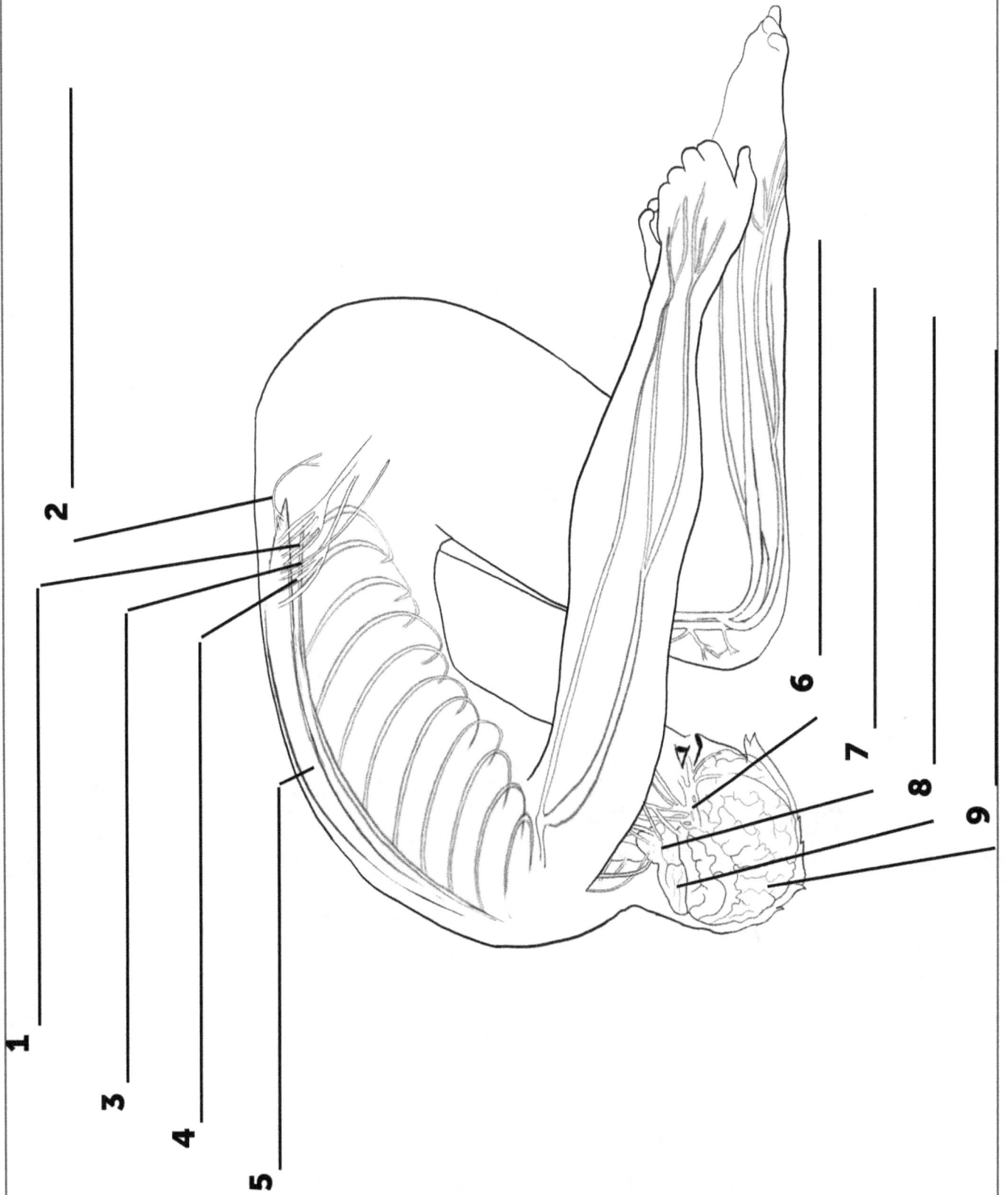

1

2

3

4

5

6

7

8

9

87. SASANGASANA

1.   PLEXO SACRO
2.   NERVIO PUDENDO
3.   OBTURADOR
4.   PLEXO LUMBAR
5.   MÉDULA ESPINAL
6.   NERVIOS CRANEALES
7.   TRONCO ENCEFÁLICO
8.   CEREBELO
9.   CEREBRO

88. PURVOTTANASANA

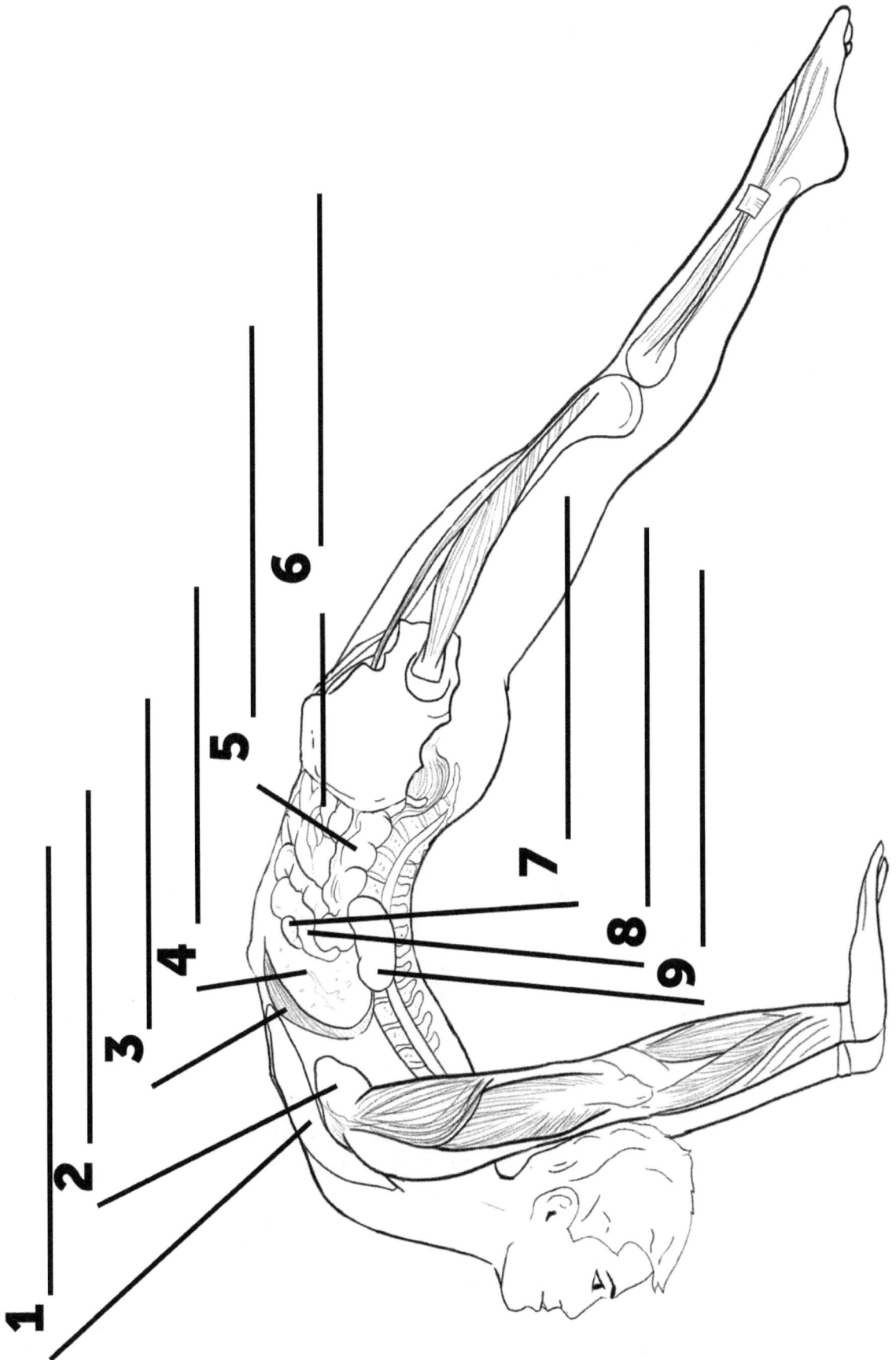

1

2

3

4

5

6

7

8

9

88. PURVOTTANASANA

1. PULMONES
2. CORAZÓN
3. DIAFRAGMA
4. HÍGADO
5. COLON ASCENDENTE
6. FOLICULOS DE INTESTINO DELGADO
7. VESÍCULA BILIAR
8. ESTÓMAGO
9. RIÑÓN

89. POSTURA DEL LOTO

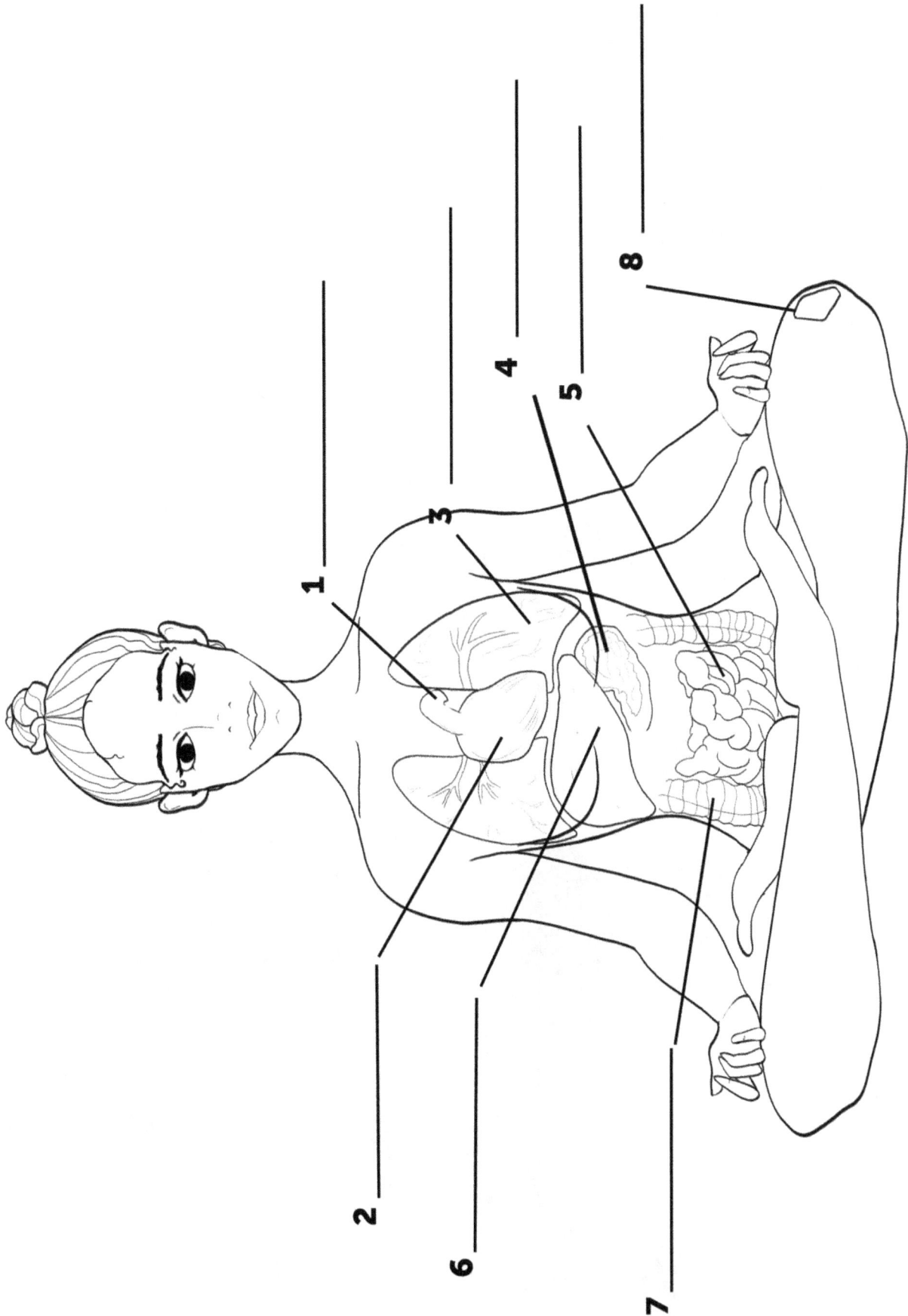

1

2

3

4

5

6

7

8

89. POSTURA DEL LOTO

1. AORTA
2. CORAZÓN
3. PULMONES
4. ESTÓMAGO
5. FOLICULOS DE INTESTINO DELGADO
6. HÍGADO
7. COLON ASCENDENTE
8. RÓTULA

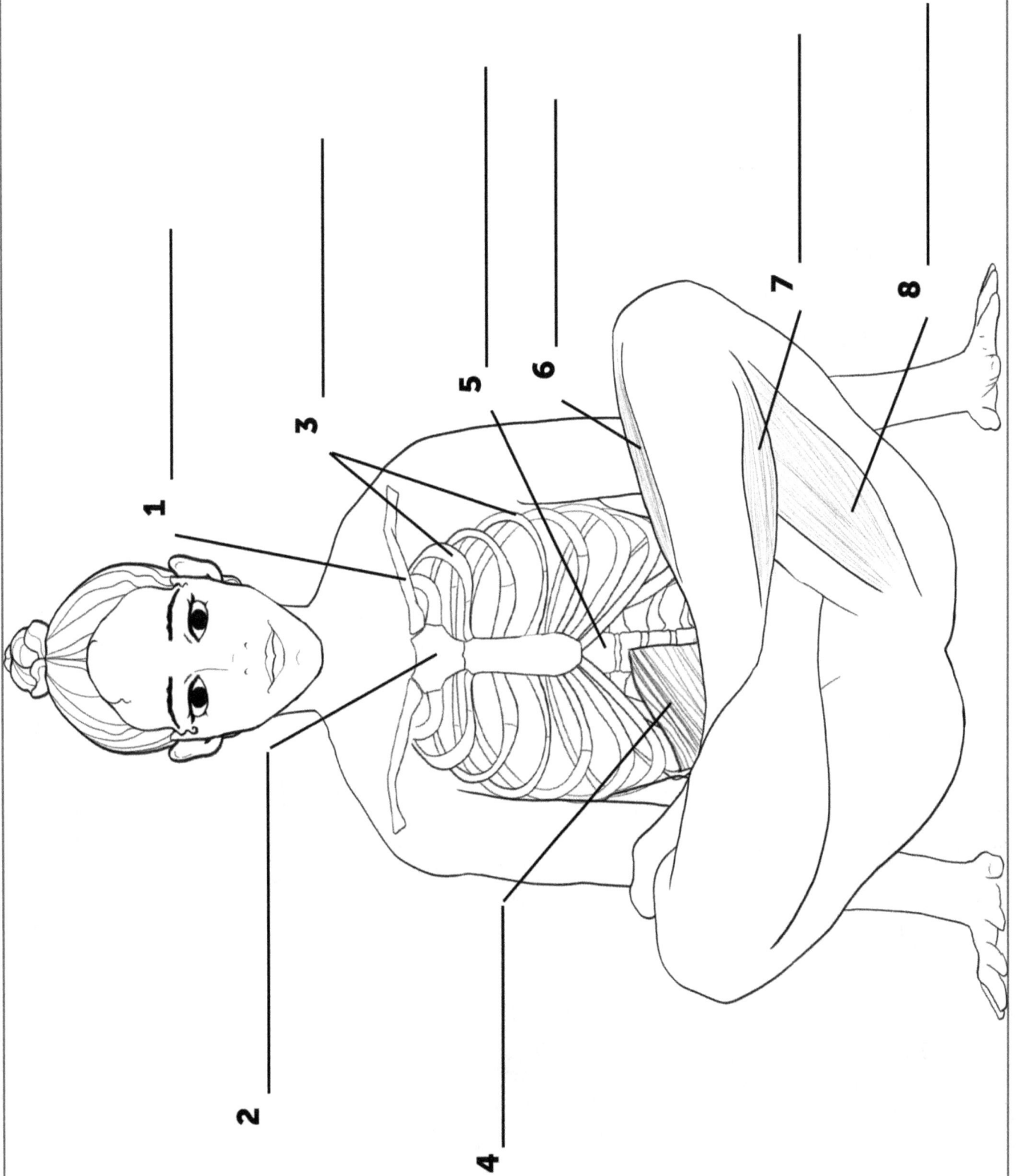

90. TOLASANA

1

2

3

4

5

6

7

8

90. TOLASANA

1. CLAVÍCULA
2. ESTERNÓN
3. COSTILLAS
4. OBLICUO INTERNO
5. COLUMNA VERTEBRAL
6. GASTROCNEMIO
7. GASTROCNEMIO
8. ISQUIOTIBIALES

91. POSTURA DEL CUERVO VOLADOR

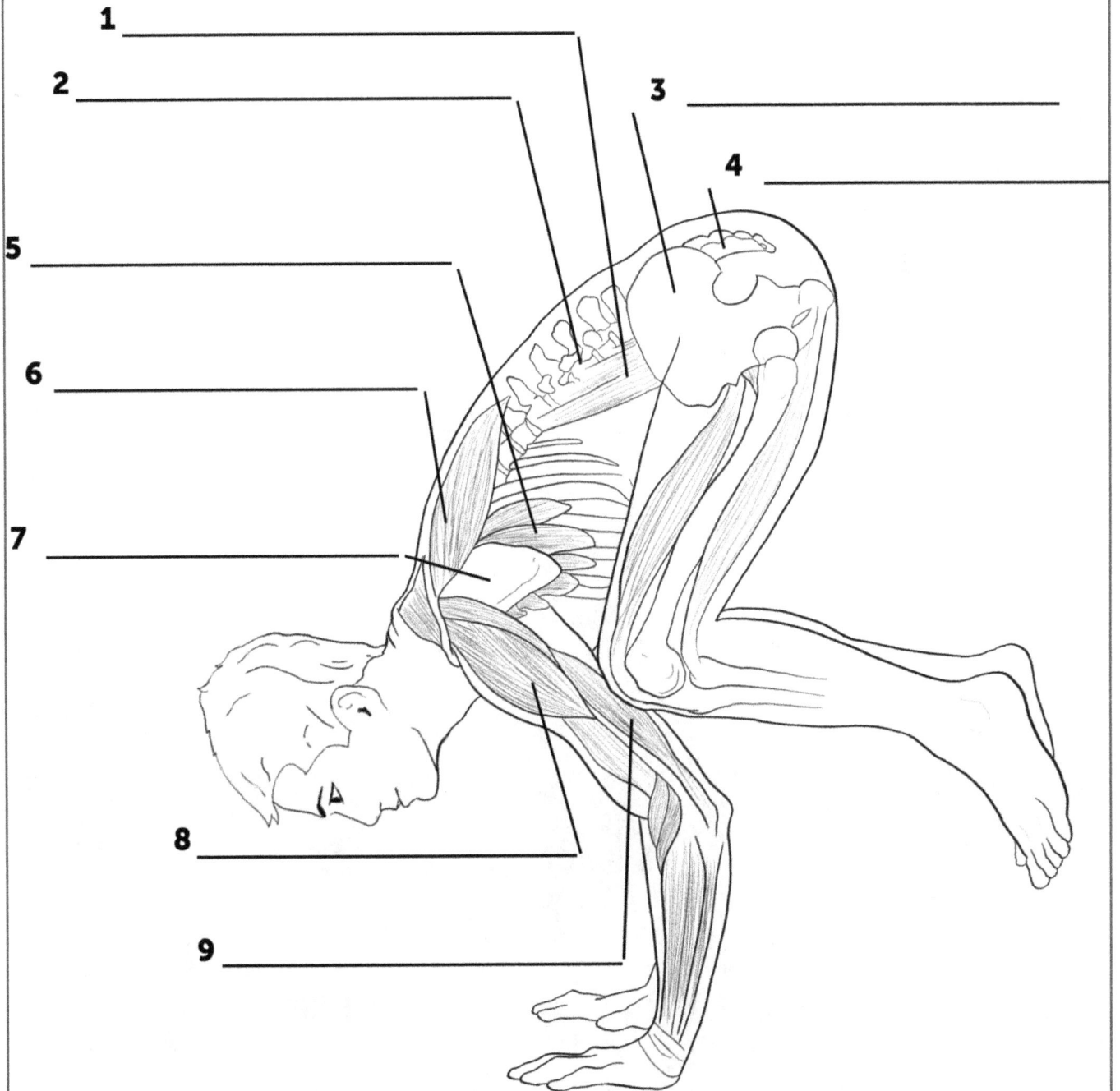

1 _____

2 _____

3 _____

4 _____

5 _____

6 _____

7 _____

8 _____

9 _____

91. POSTURA DEL CUERVO VOLADOR

1. MÚSCULO PSOAS MAYOR
2. COLUMNA VERTEBRAL
3. PELVIS
4. SACRO
5. SERRATO ANTERIOR
6. TRAPECIO
7. ESCÁPULA
8. DELTOIDES
9. TRÍCEPS BRAQUIAL

92. CHATURANGA DANDASANA

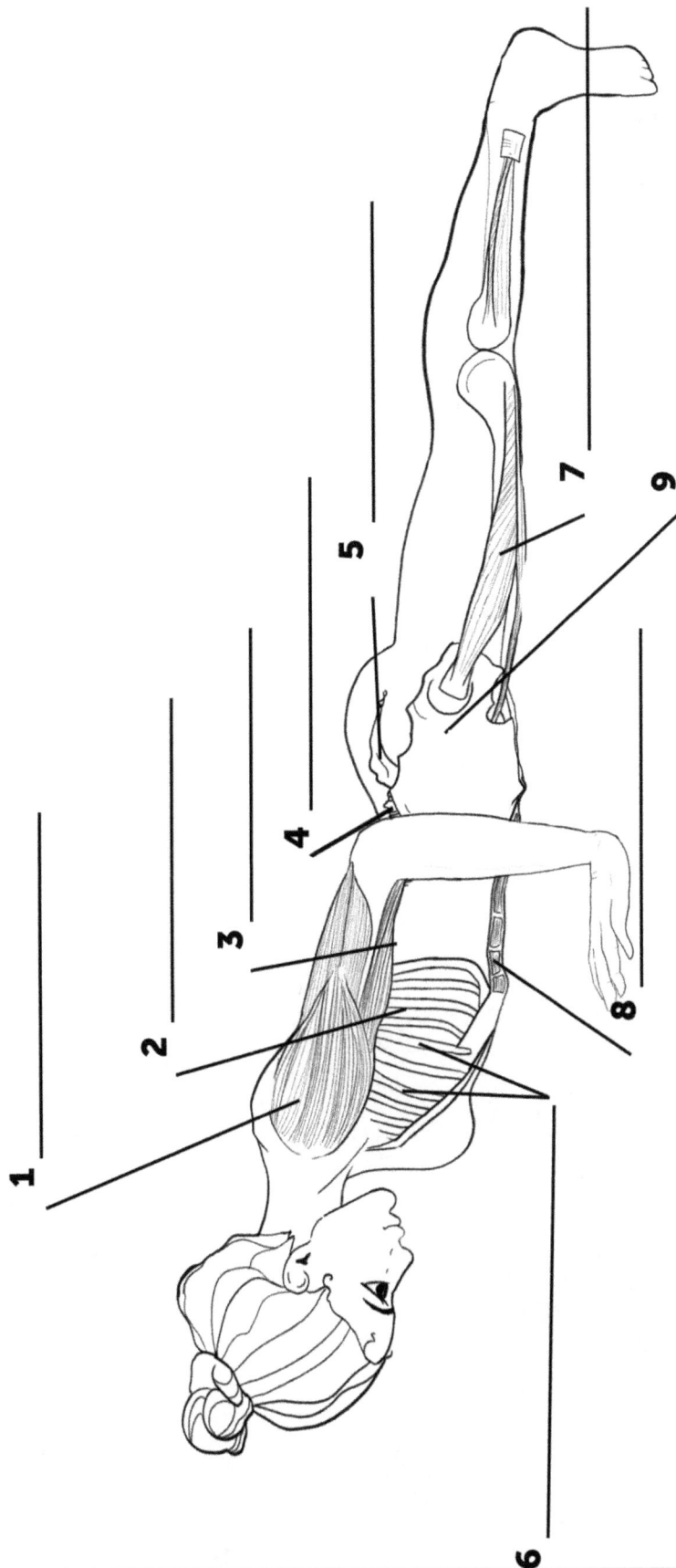

92. CHATURANGA DANDASANA

1. DELTOIDES
2. COSTILLAS
3. BÍCEPS BRAQUIAL
4. COLUMNA VERTEBRAL
5. SACRO
6. COSTILLAS
7. RECTO FEMORAL
8. RECTO ABDOMINAL
9. PELVIS

93. PARSVA BAKASANA

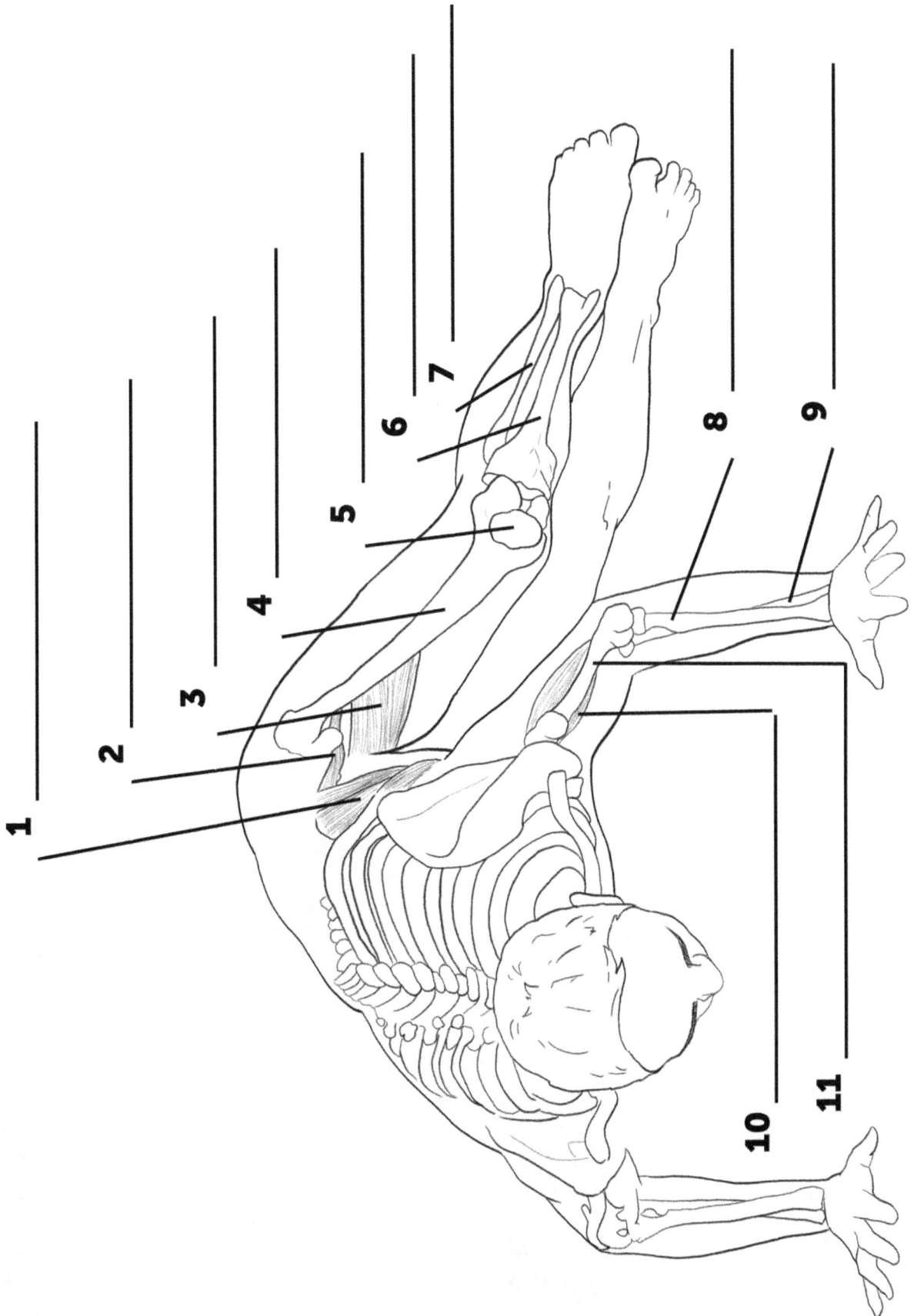

93. PARSVA BAKASANA

1. OBLICUO EXTERNO
2. PECTÍNEO
3. ADUCTOR CORTO
4. FÉMUR
5. RÓTULA
6. TIBIA
7. PERONÉ
8. RADIO
9. CÚBITO
10. TRÍCEPS BRAQUIAL
11. HÚMERO

94. ARDHA NAVASANA

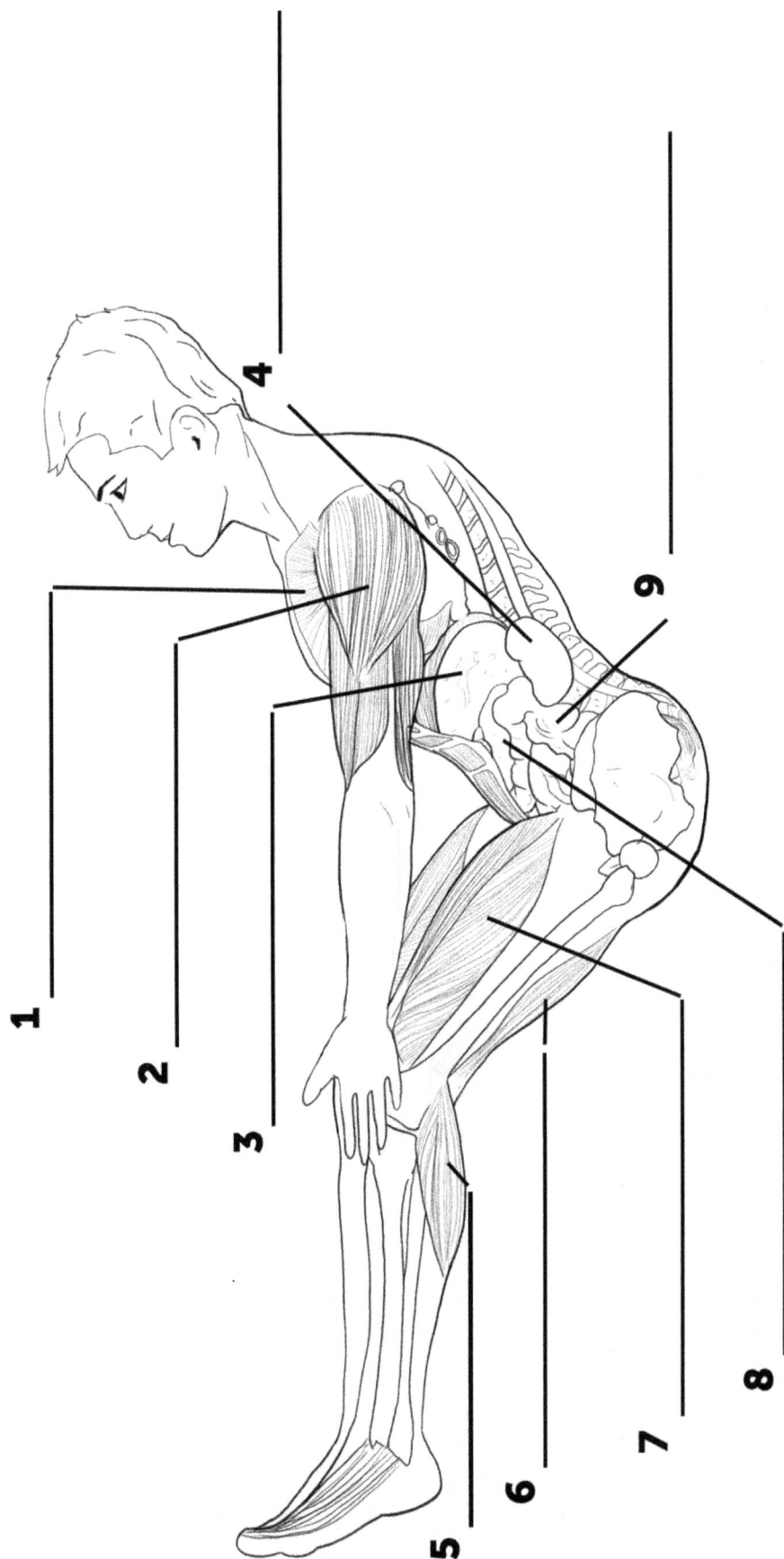

1

2

3

4

5

6

7

8

9

94. ARDHA NAVASANA

1. PECTORAL MAYOR
2. DELTOIDES
3. HÍGADO
4. RIÑÓN
5. GASTROCNEMIO
6. ISQUIOTIBIALES
7. CUADRÍCEPS
8. ESTÓMAGO
9. COLON ASCENDENTE

95. PARIPURNA NAVASANA

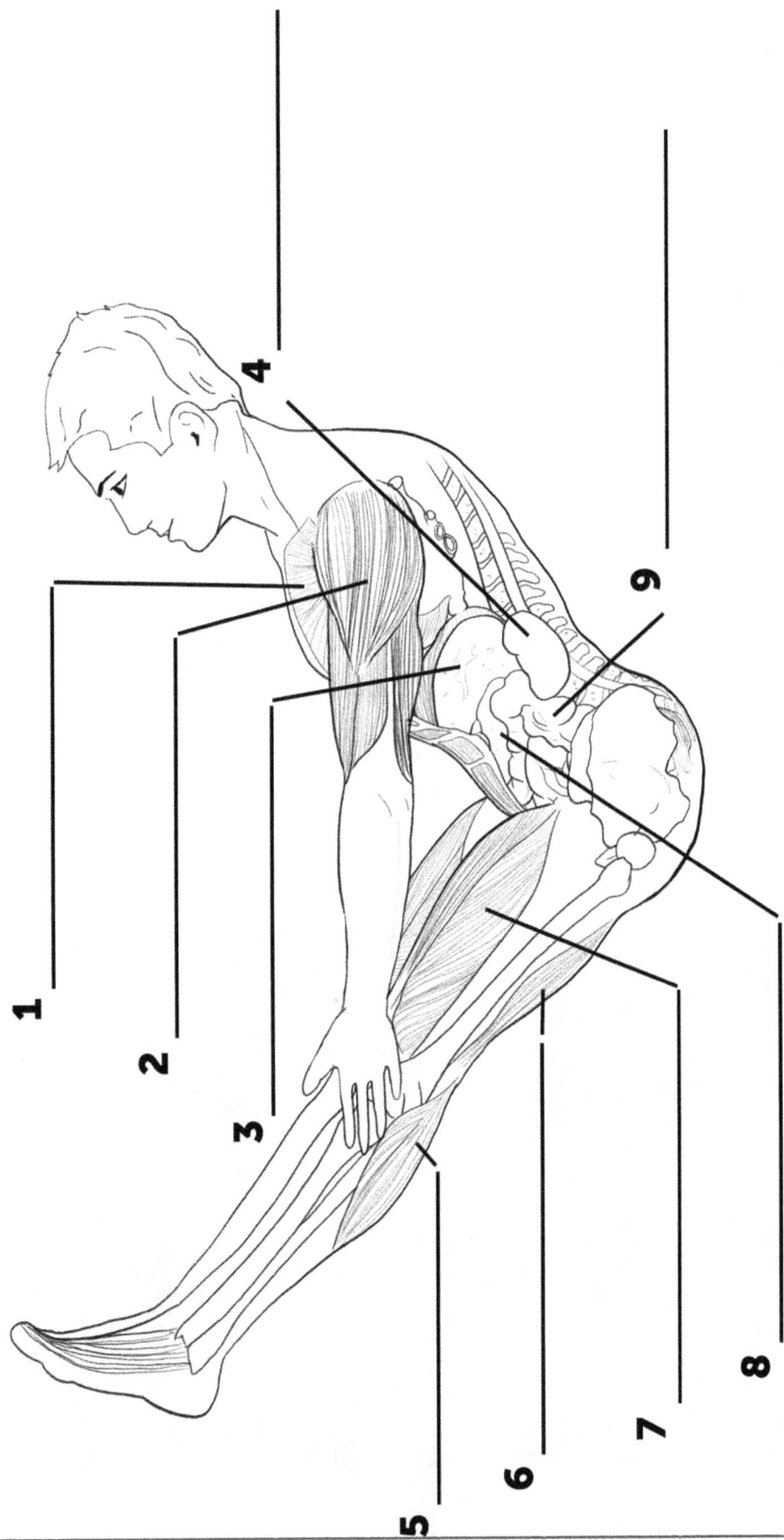

1

2

3

4

5

6

7

8

9

95. PARIPURNA NAVASANA

1. PECTORAL MAYOR
2. DELTOIDES
3. HÍGADO
4. RIÑÓN
5. GASTROCNEMIO
6. ISQUIOTIBIALES
7. CUADRÍCEPS
8. ESTÓMAGO
9. COLON ASCENDENTE

96. MATSYASANA

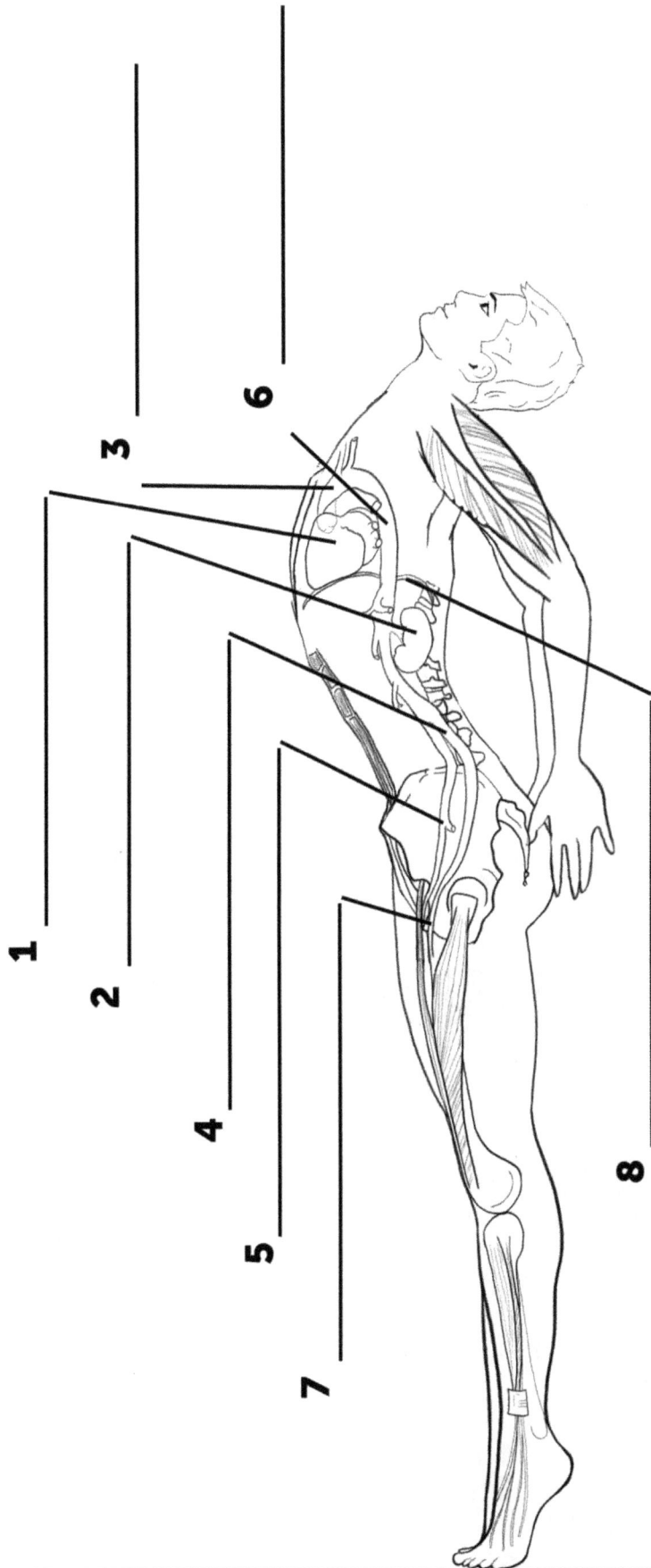

96. MATSYASANA

1. CORAZÓN
2. RIÑÓN
3. AORTA TORÁCICA ASCENDENTE
4. AORTA ABDOMINAL
5. ARTERIA ILIACA COMÚN
6. AORTA TORÁCICA DESCENDENTE
7. ARTERIA FEMORAL
8. DIAFRAGMA

97. SHIRSASANA

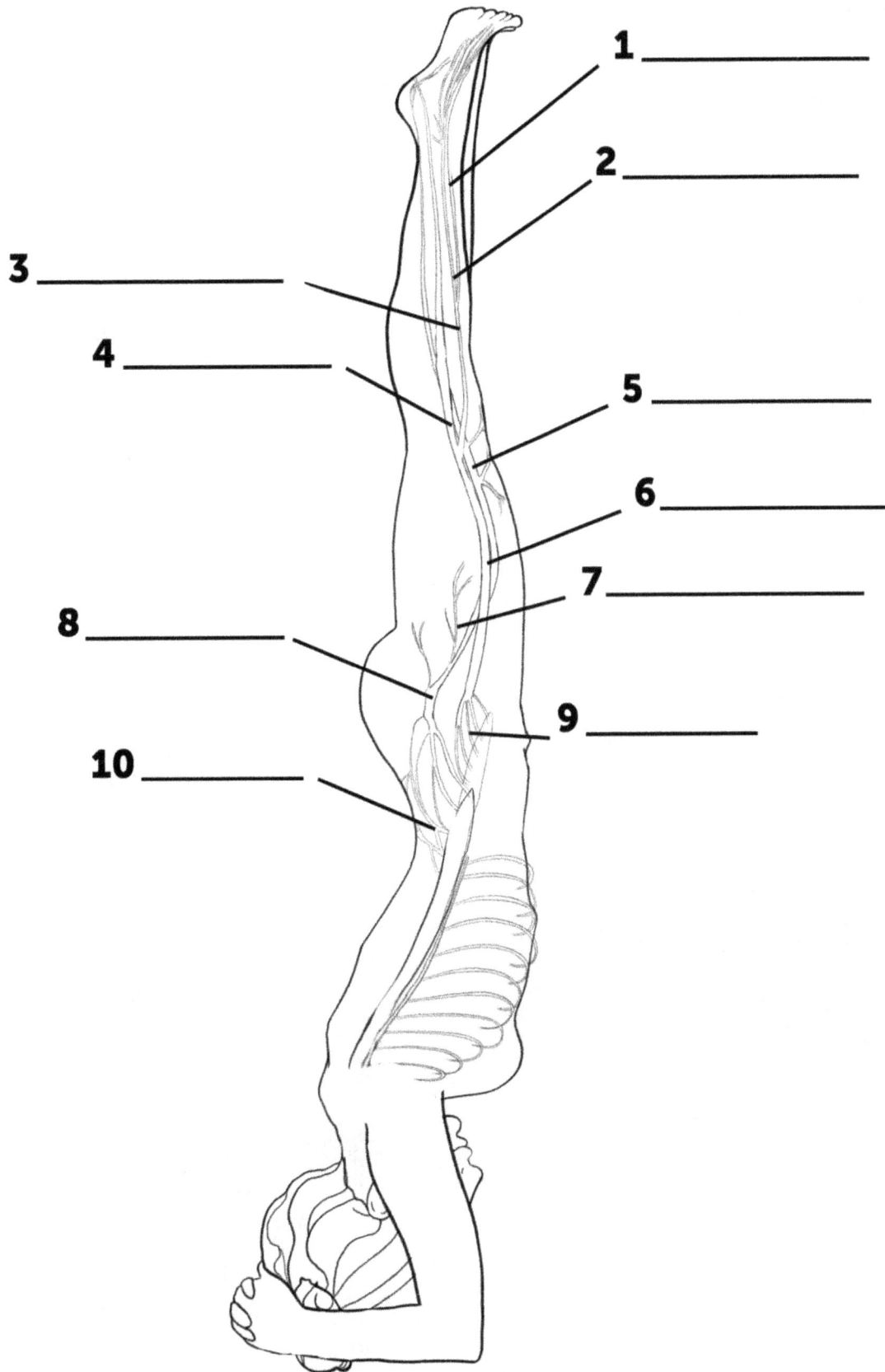

1 _____

2 _____

3 _____

4 _____

5 _____

6 _____

7 _____

8 _____

9 _____

10 _____

97. SHIRSASANA

1. PERONEO SUPERFICIAL

2. PERONEO PROFUNDO

3. PERONEO COMÚN

4. TIBIAL

5. SAFENA

6. CIÁTICO

7. RAMAS MUSCULARES DE FEMORAL

8. FEMORAL

9. PLEXO SACRO

10. PLEXO LUMBAR

98. SALAMBA SARVANGASANA

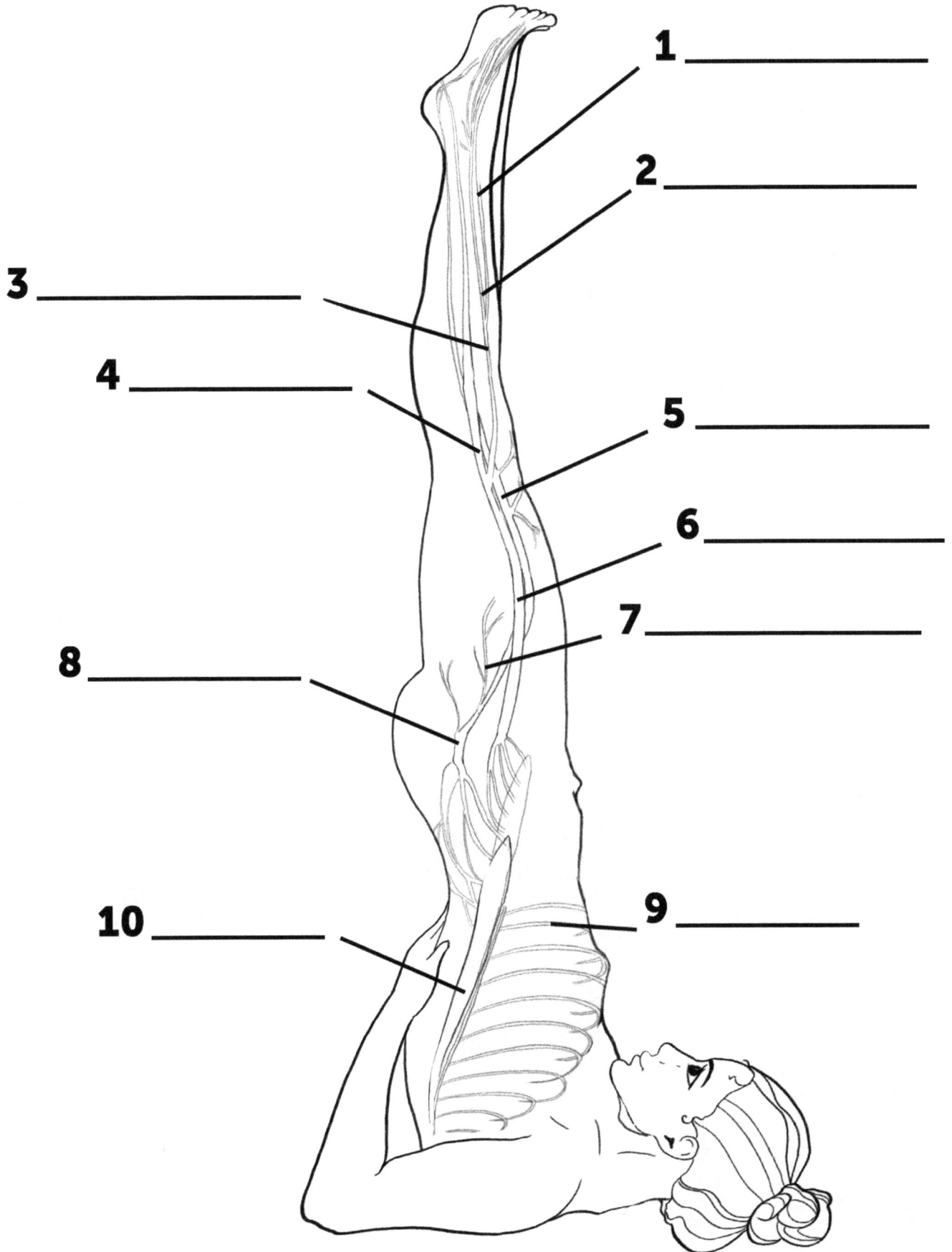

1 _____

2 _____

3 _____

4 _____

5 _____

6 _____

7 _____

8 _____

9 _____

10 _____

98. SALAMBA SARVANGASANA

1. PERONEO SUPERFICIAL
2. PERONEO PROFUNDO
3. PERONEO COMÚN
4. TIBIAL
5. SAFENA
6. CIÁTICO
7. RAMAS MUSCULARES DE FEMORAL
8. FEMORAL
9. INTERCOSTALES
10. MÉDULA ESPINAL

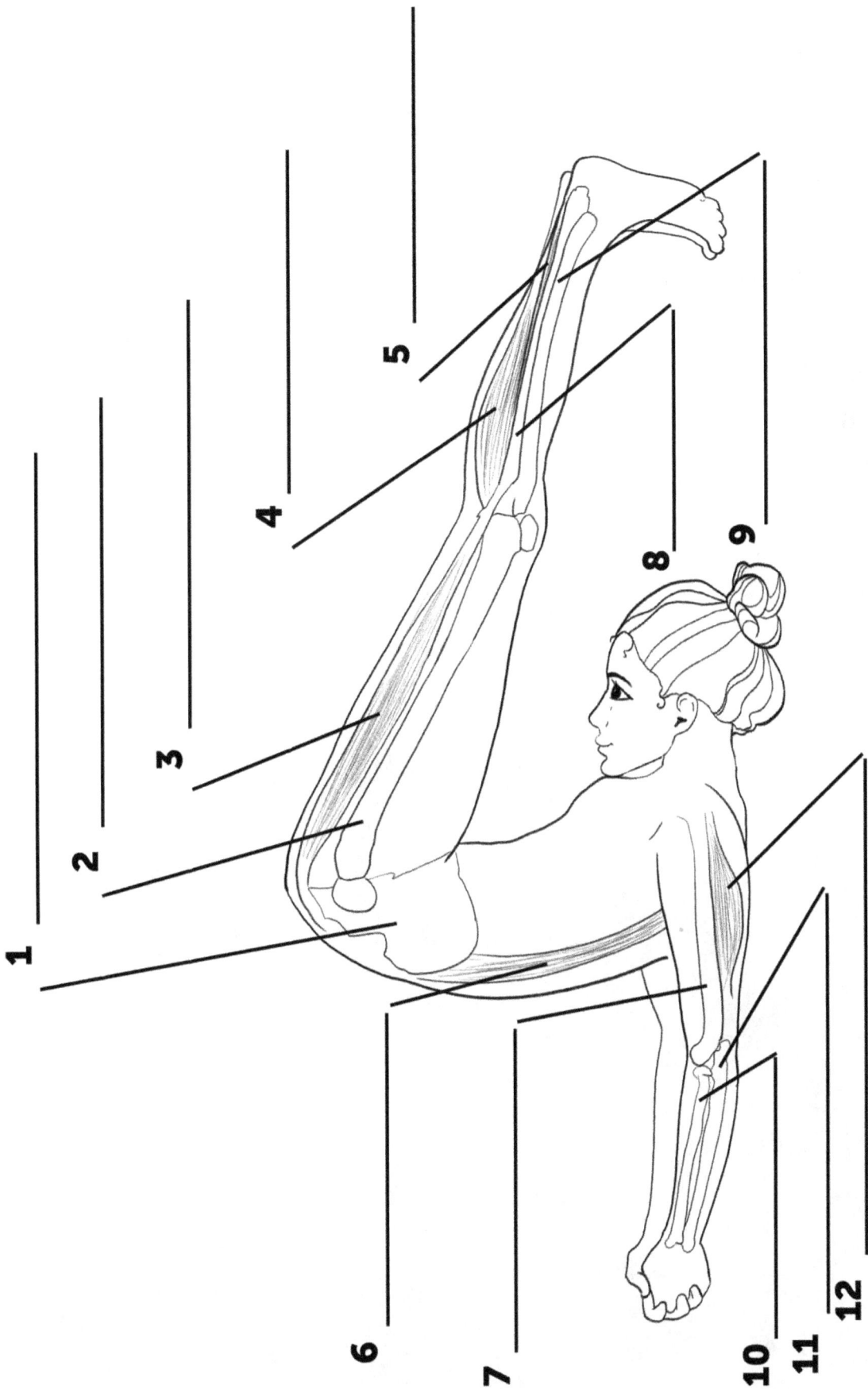

99. HALASANA

1

2

3

4

5

6

7

8

9

10

11

12

99. HALASANA

1.   PELVIS
2.   FÉMUR
3.   ISQUIOTIBIALES
4.   GASTROCNEMIO
5.   SÓLEO
6.   ERECTOR DE LA COLUMNA
7.   HÚMERO
8.   PERONÉ
9.   TIBIA
10. RADIO
11. CÚBITO
12. TRÍCEPS BRAQUIAL

100. KARNAPIDASANA

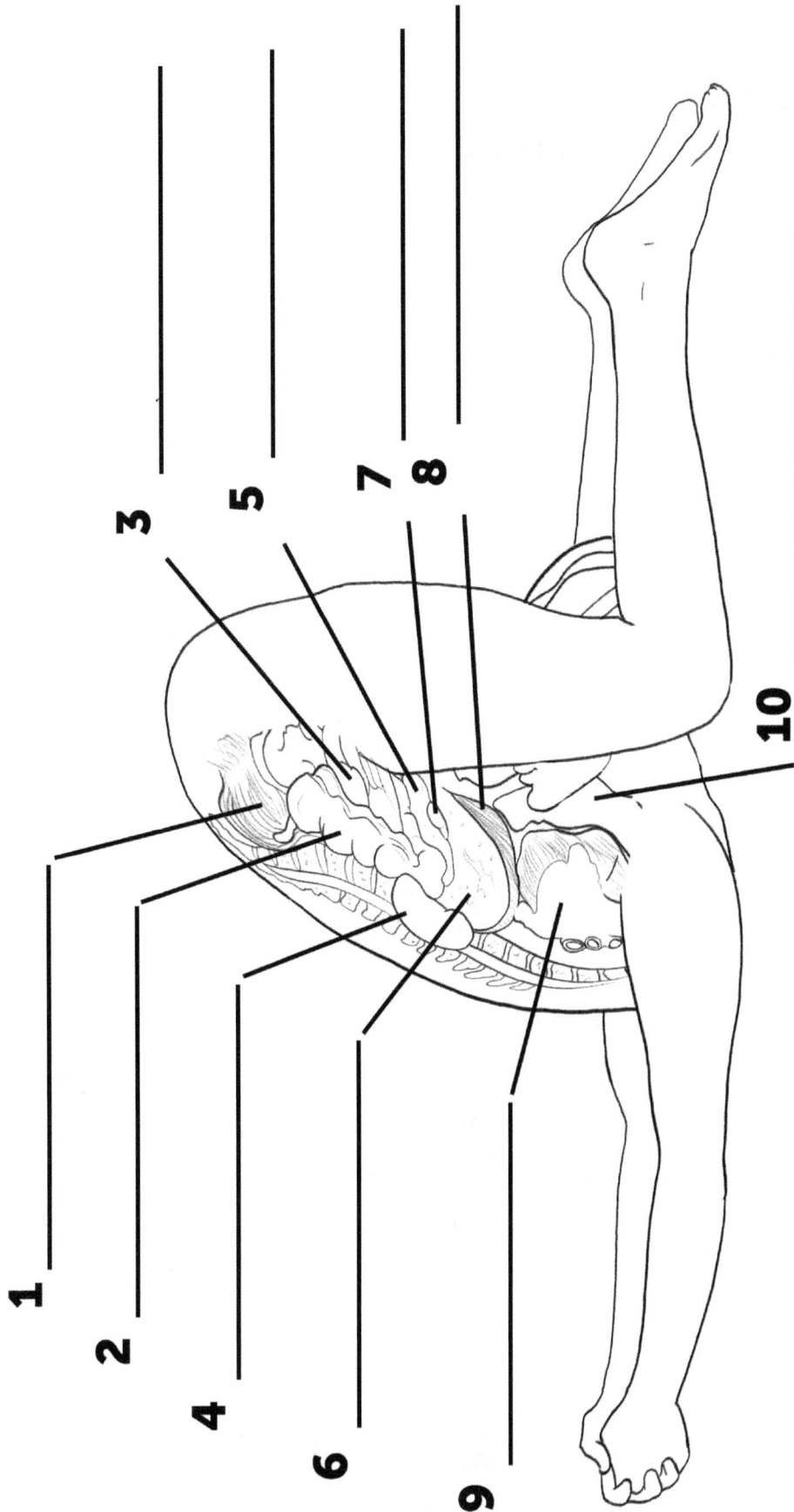

100. KARNAPIDASANA

1. RECTO
2. COLON ASCENDENTE
3. FOLICULOS DE INTESTINO DELGADO
4. RIÑÓN
5. ESTÓMAGO
6. HÍGADO
7. VESÍCULA BILIAR
8. DIAFRAGMA
9. CORAZÓN
10. PULMONES

101. POSTURA DE MEDIA LUNA

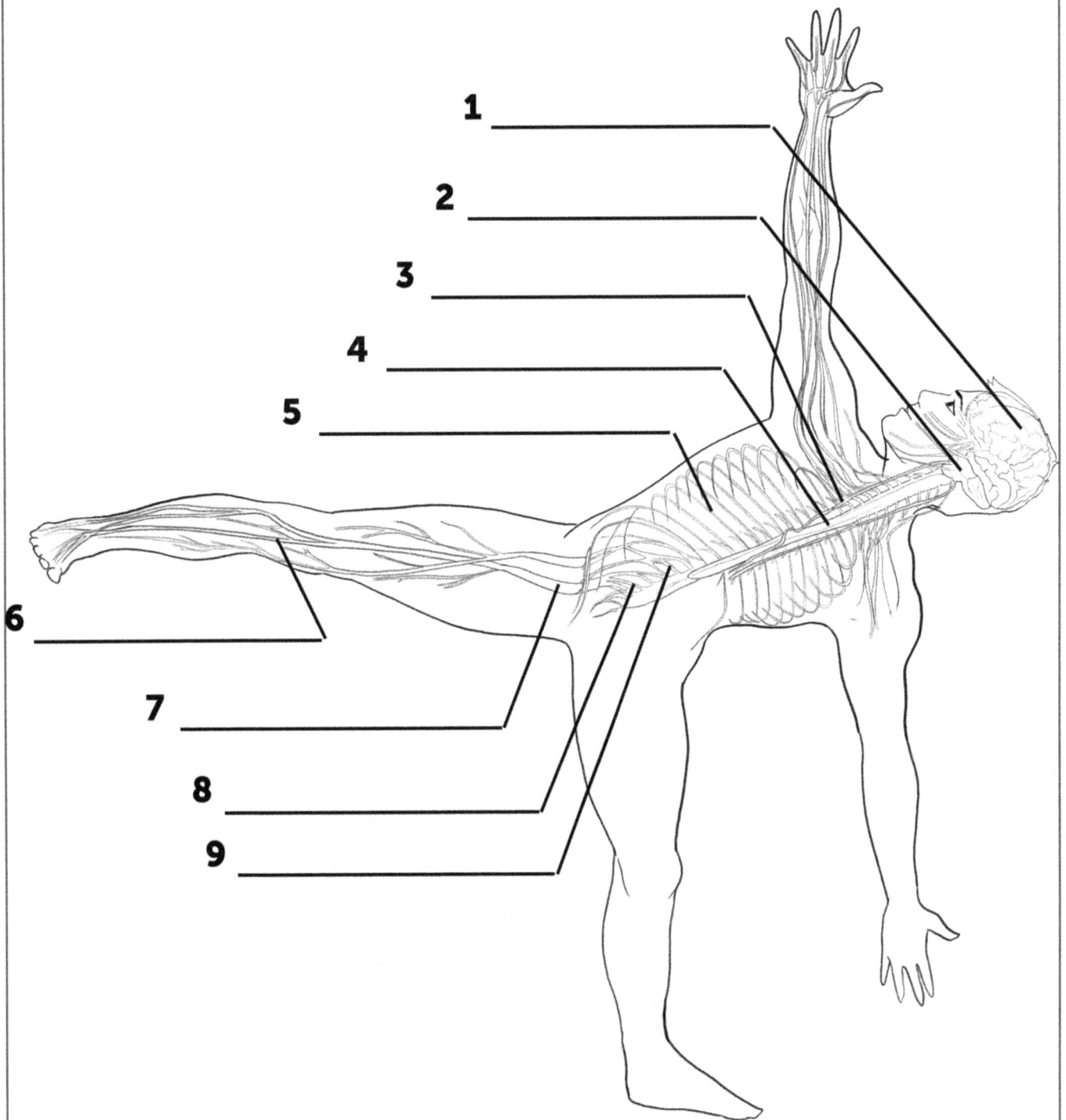

1 _____

2 _____

3 _____

4 _____

5 _____

6 _____

7 _____

8 _____

9 _____

101. POSTURA DE MEDIA LUNA

1.   CEREBRO

2.   TRONCO ENCEFÁLICO

3.   PLEXO BRAQUIAL

4.   MÉDULA ESPINAL

5.   INTERCOSTALES

6.   TIBIAL

7.   CIÁTICO

8.   PLEXO SACRO

9.   PLEXO LUMBAR

102. PARIVRTTA SURYA YANTRASANA

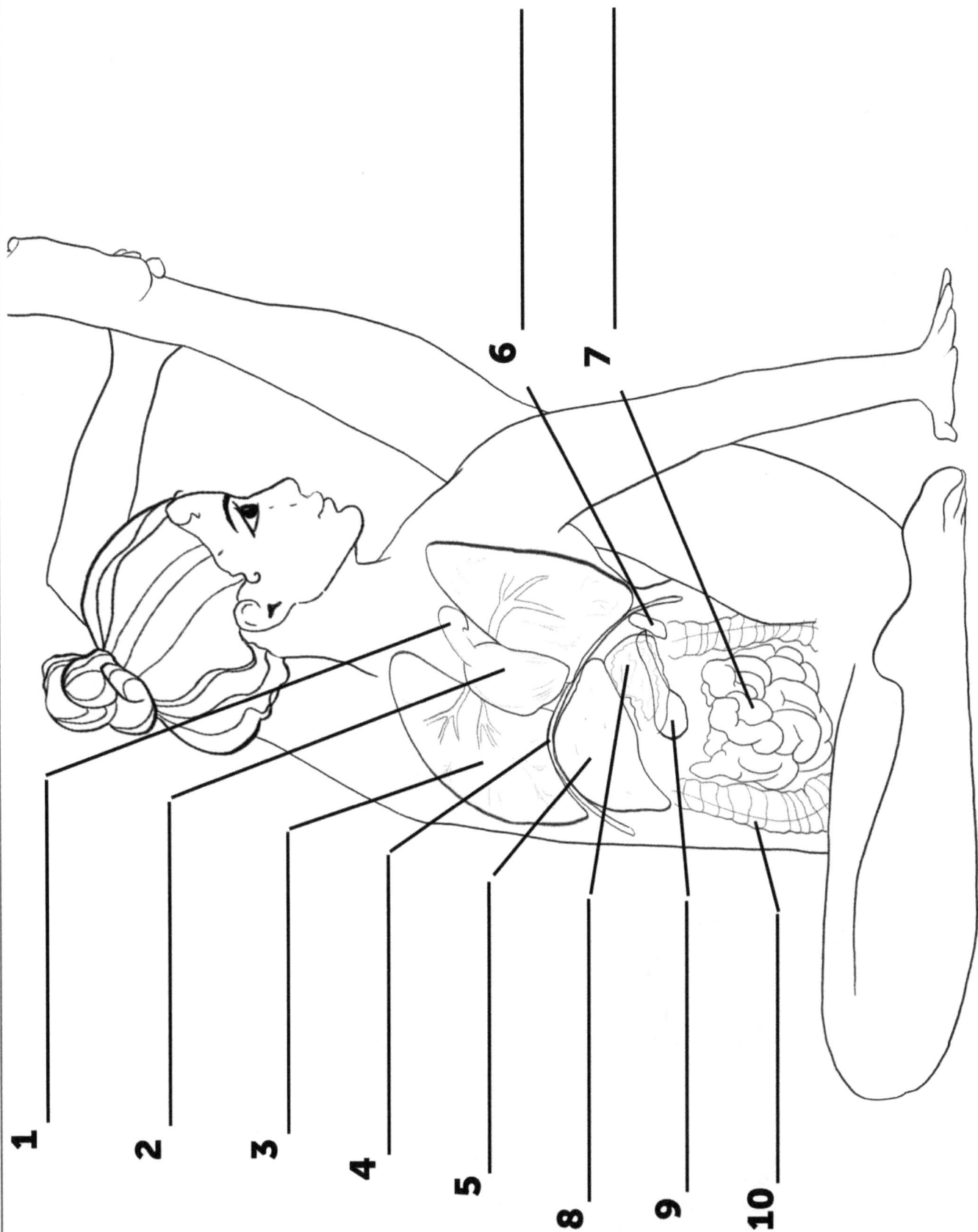

1

2

3

4

5

6

7

8

9

10

102. PARIVRTTA SURYA YANTRASANA

1. AORTA
2. CORAZÓN
3. PULMONES
4. DIAFRAGMA
5. HÍGADO
6. BAZO
7. FOLICULOS DE INTESTINO DELGADO
8. ESTÓMAGO
9. PÁNCREAS
10. COLON ASCENDENTE

103. PARIVRTTA JANU SIRSASANA IN SANSKRIT

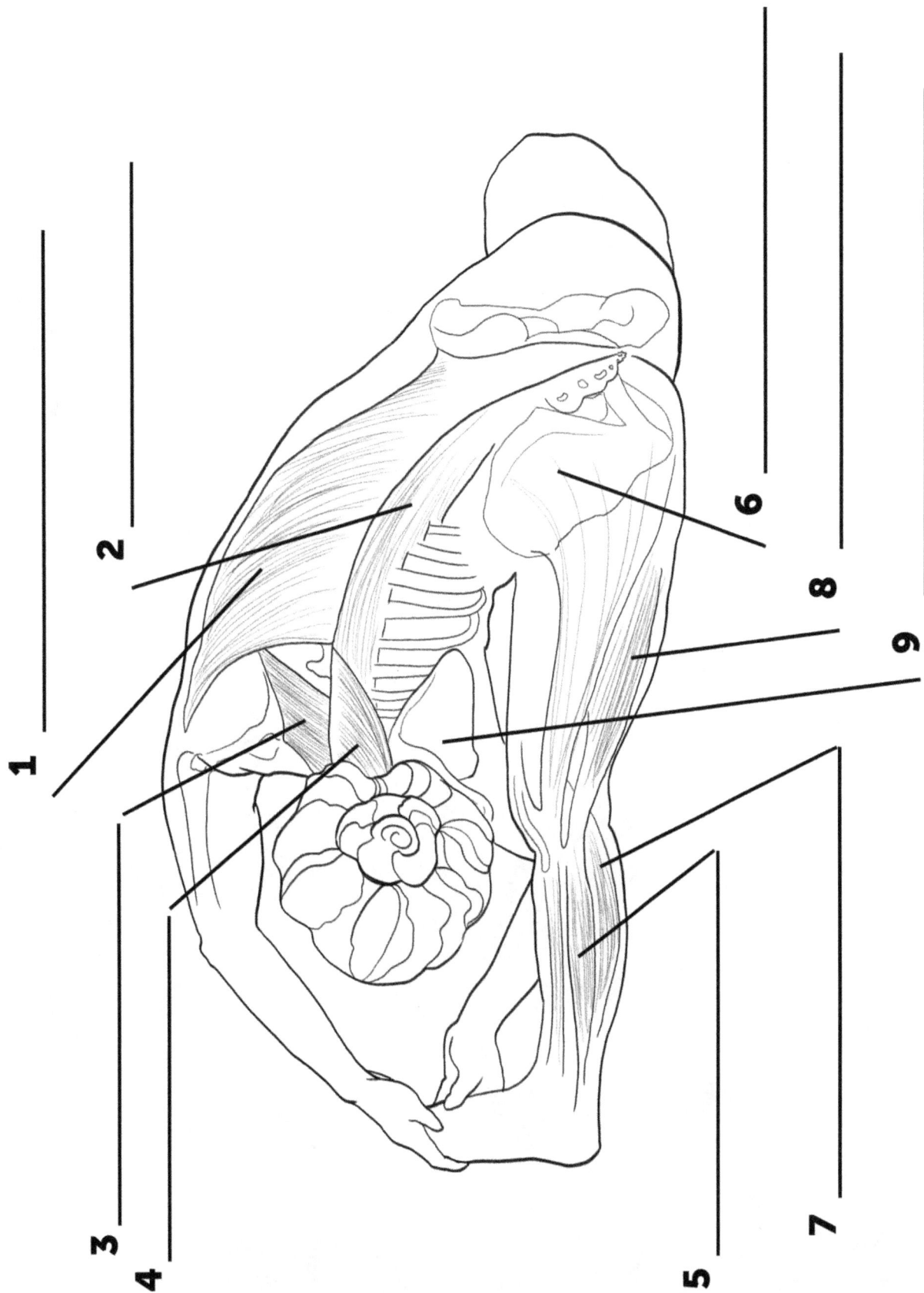

1

2

3

4

5

6

7

8

9

103. PARIVRTTA JANU SIRSASANA IN SANSKRIT

1. LATISSIMUS DORSI

2. ERECTOR DE LA COLUMNA

3. ROMBOIDES

4. TRAPECIO

5. SÓLEO

6. PELVIS

7. GASTROCNEMIO

8. ISQUIOTIBIALES

9. ESCÁPULA

104. POSTURA DIVIDIDA DE PIE

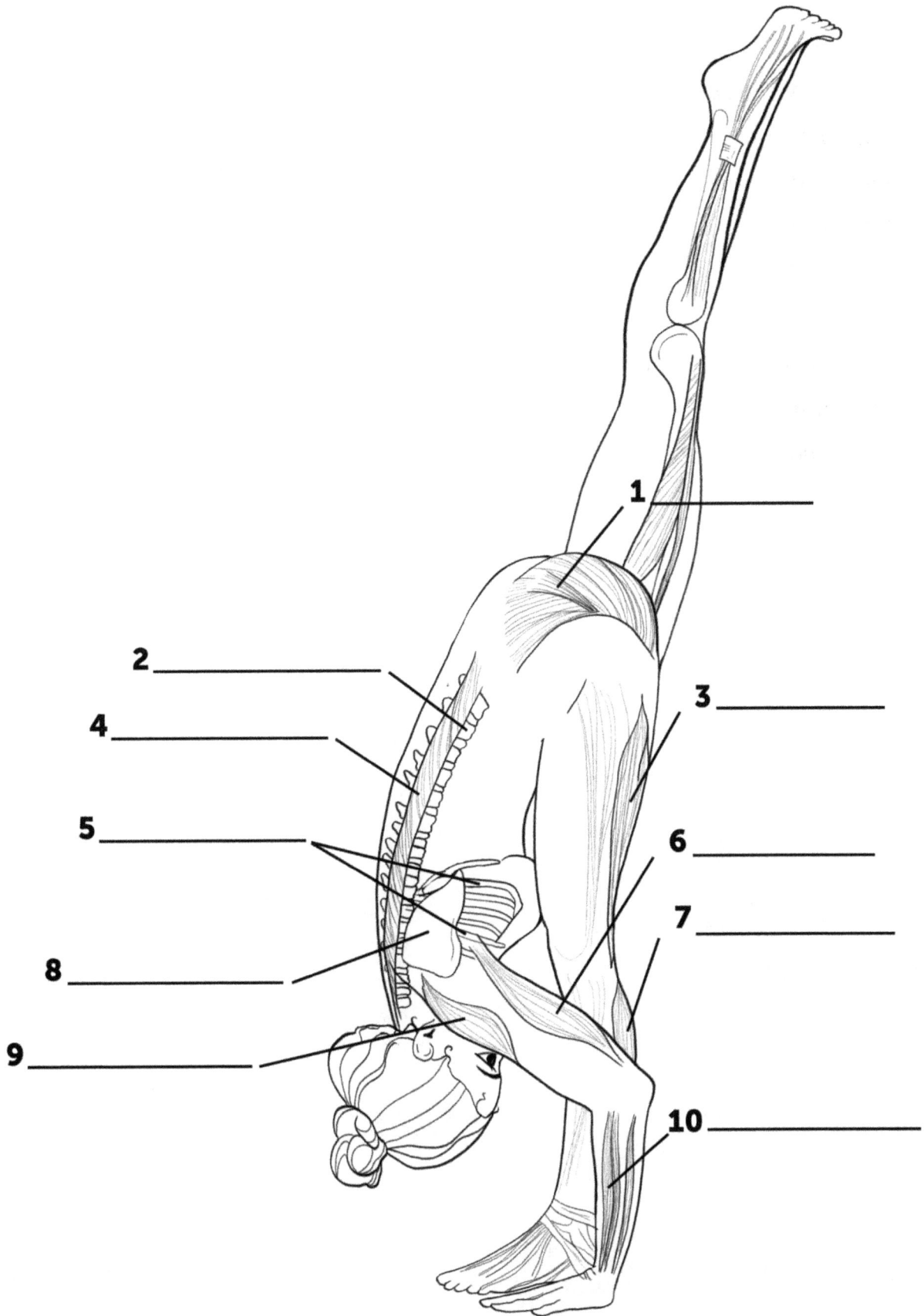

104. POSTURA DIVIDIDA DE PIE

1. PIRIFORME
2. COLUMNA VERTEBRAL
3. ISQUIOTIBIALES
4. ERECTOR DE LA COLUMNA
5. COSTILLAS
6. TRÍCEPS BRAQUIAL
7. GASTROCNEMIO
8. ESCÁPULA
9. DELTOIDES
10. PRONADORES

105. AKARNA DHANURASANA

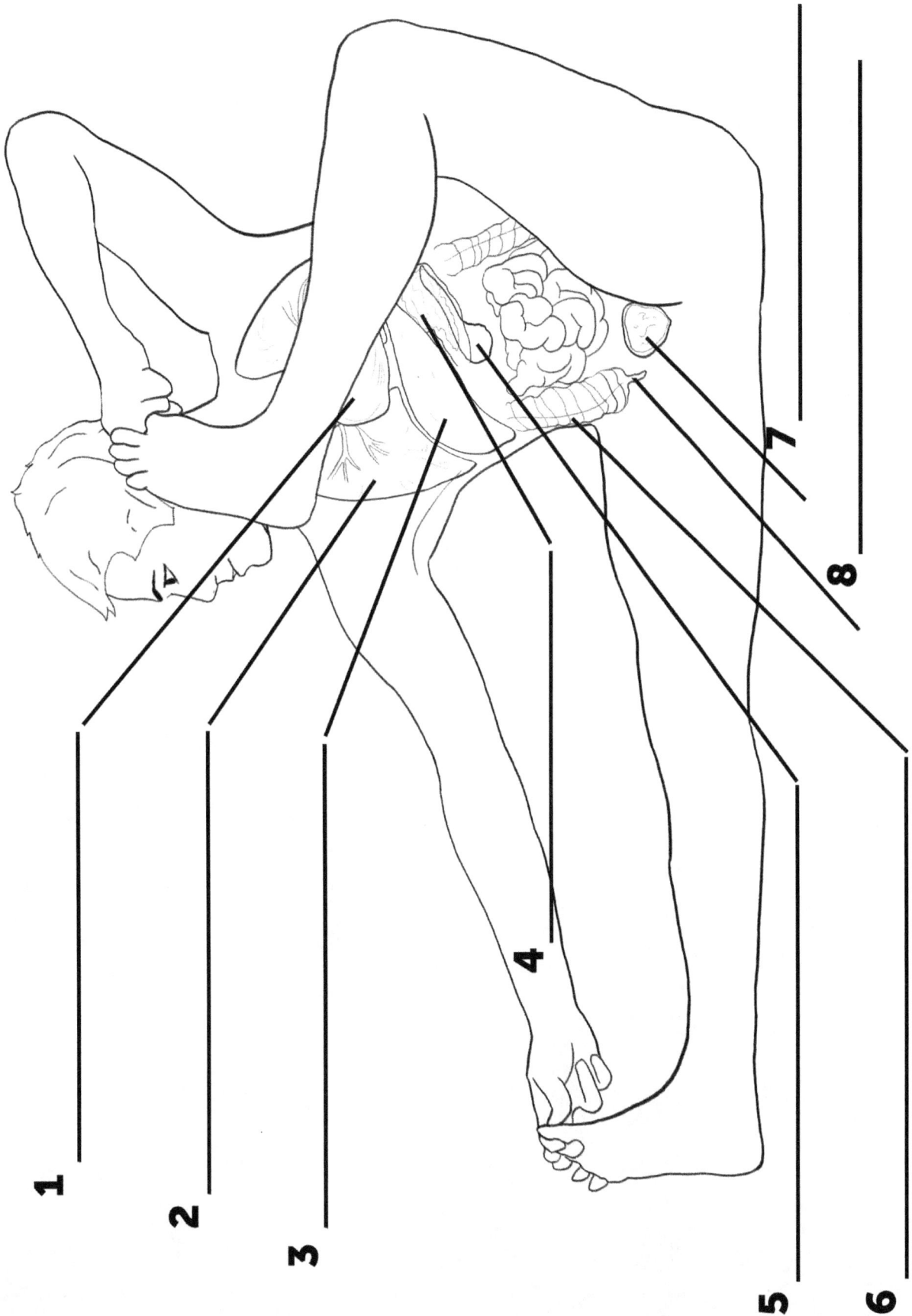

1

2

3

4

5

6

7

8

105. AKARNA DHANURASANA

1. CORAZÓN
2. PULMONES
3. HÍGADO
4. ESTÓMAGO
5. PÁNCREAS
6. COLON ASCENDENTE
7. VEJIGA URINARIA
8. APÉNDICE

106. ADHO MUKHA VRKSASANA

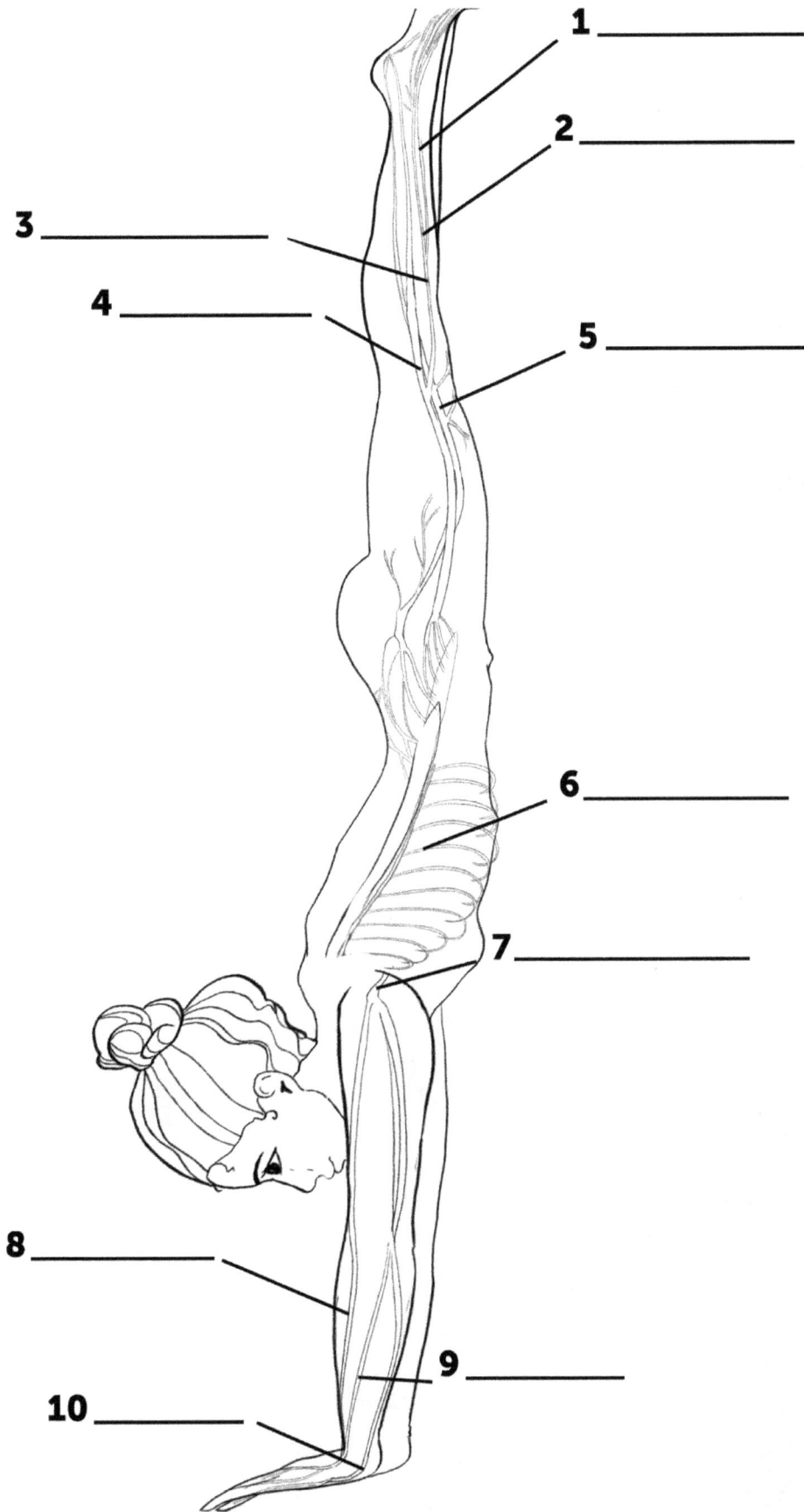

1 _____

2 _____

3 _____

4 _____

5 _____

6 _____

7 _____

8 _____

9 _____

10 _____

106. ADHO MUKHA VRKSASANA

1. PERONEO SUPERFICIAL
2. PERONEO PROFUNDO
3. PERONEO COMÚN
4. TIBIAL
5. SAFENA
6. INTERCOSTALES
7. PLEXO BRAQUIAL
8. RADIAL
9. MEDIANA
10. ULNAR

107. EKA HASTA BHUJASANA

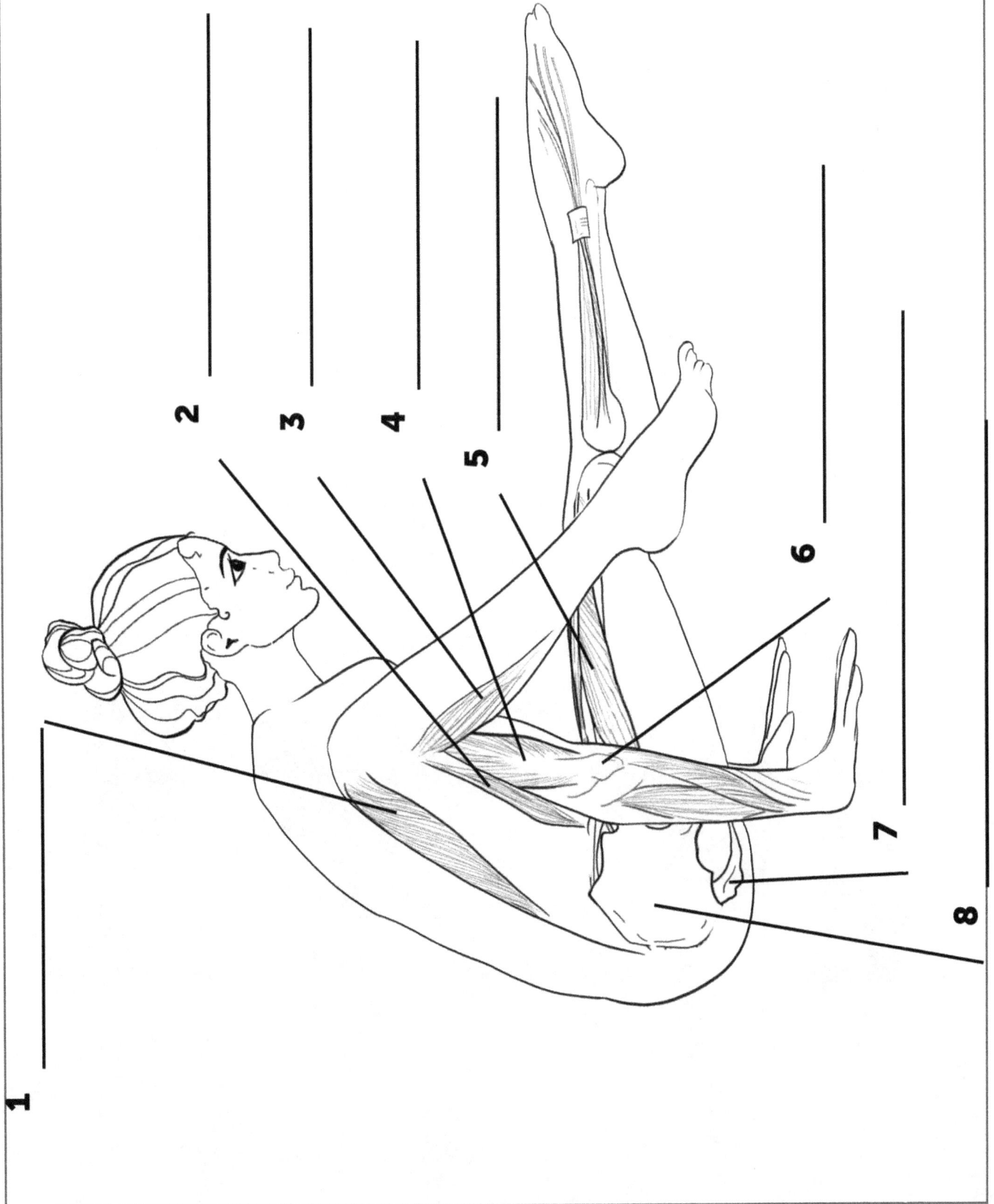

1

2

3

4

5

6

7

8

107. EKA HASTA BHUJASANA

1. RECTO FEMORAL
2. ISQUIOTIBIALES
3. GASTROCNEMIO
4. TRÍCEPS BRAQUIAL
5. CUADRÍCEPS
6. CODO
7. SACRO
8. PELVIS

POSTURAS DE YOGA PARA EXPERTOS

108. PURNA MATSYENDRASANA

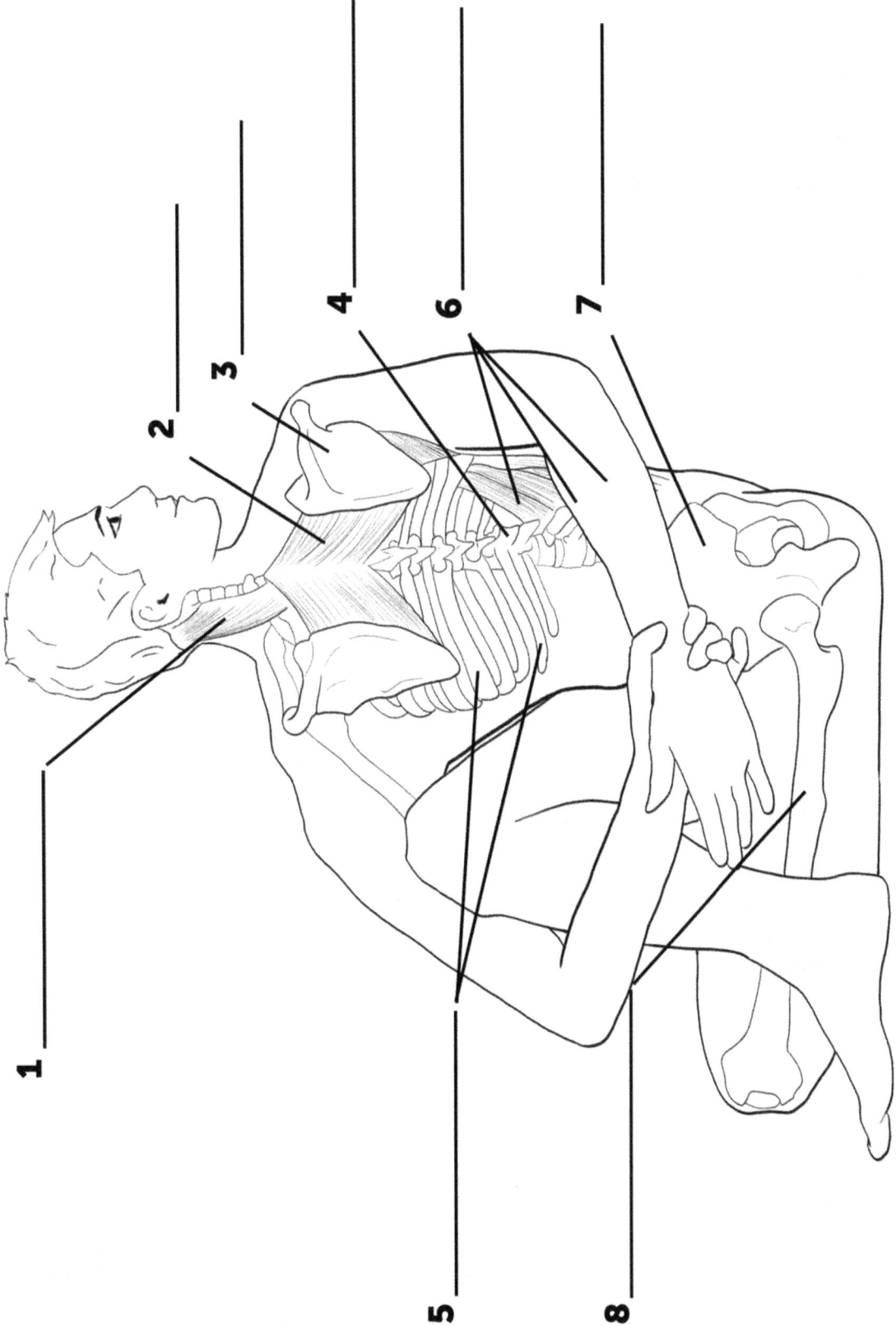

108. PURNA MATSYENDRASANA

1. MÚSCULO ESPLENIO DE LA CABEZA

2. ROMBOIDES

3. ESCÁPULA

4. COLUMNA VERTEBRAL

5. COSTILLAS

6. ERECTOR DE LA COLUMNA

7. PELVIS

8. FÉMUR

109. POSTURA DEL CUERVO VOLADOR

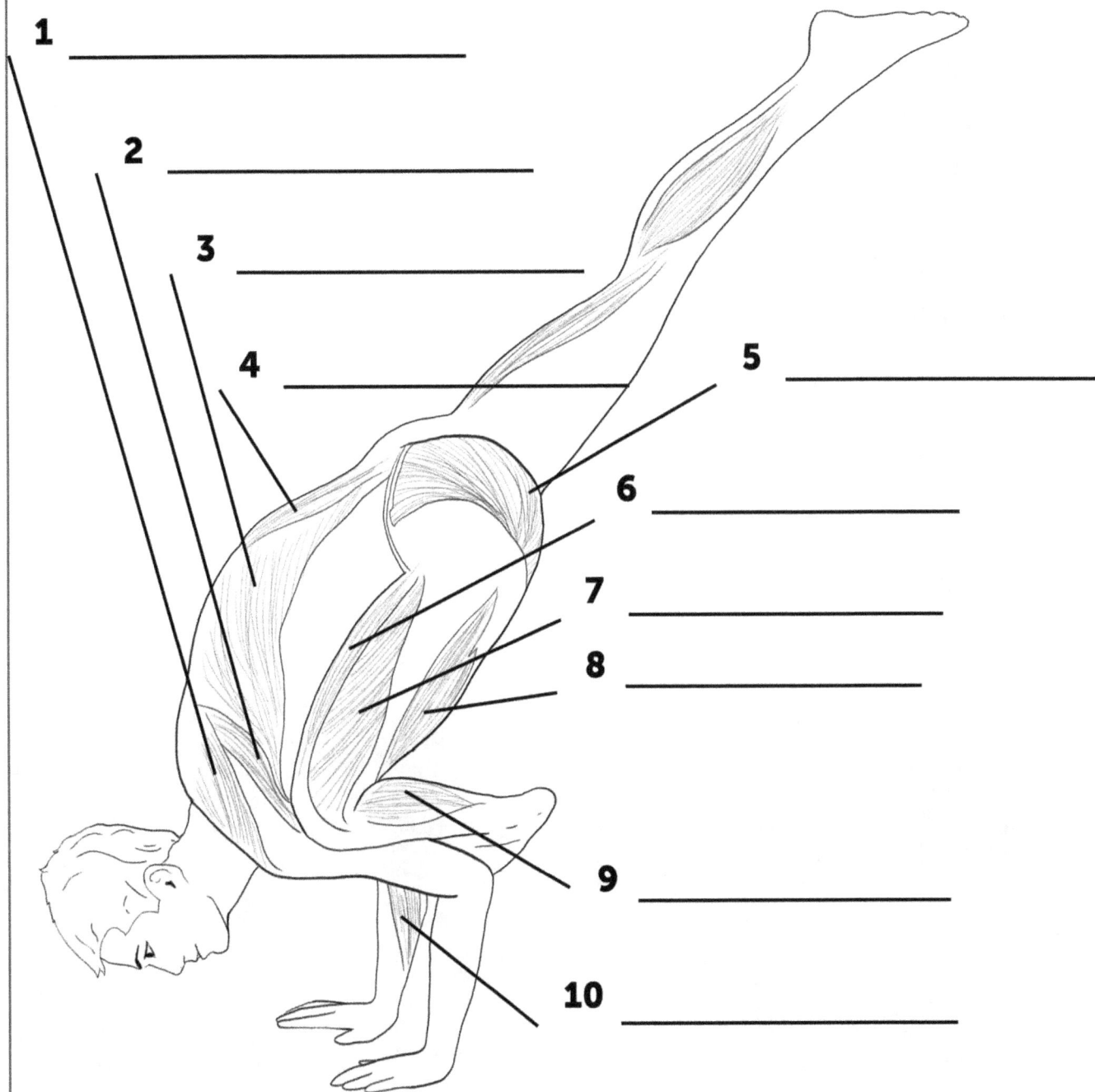

1 _____

2 _____

3 _____

4 _____

5 _____

6 _____

7 _____

8 _____

9 _____

10 _____

109. POSTURA DEL CUERVO VOLADOR

1. DELTOIDES
2. TRÍCEPS BRAQUIAL
3. LATISSIMUS DORSI
4. ERECTOR DE LA COLUMNA
5. MÚSCULO GLÚTEO MAYOR
6. RECTO FEMORAL
7. MÚSCULO VASTO LATERAL
8. ISQUIOTIBIALES
9. GASTROCNEMIO
10. PRONADORES

110. POSTURA DEL ESCORPIÓN

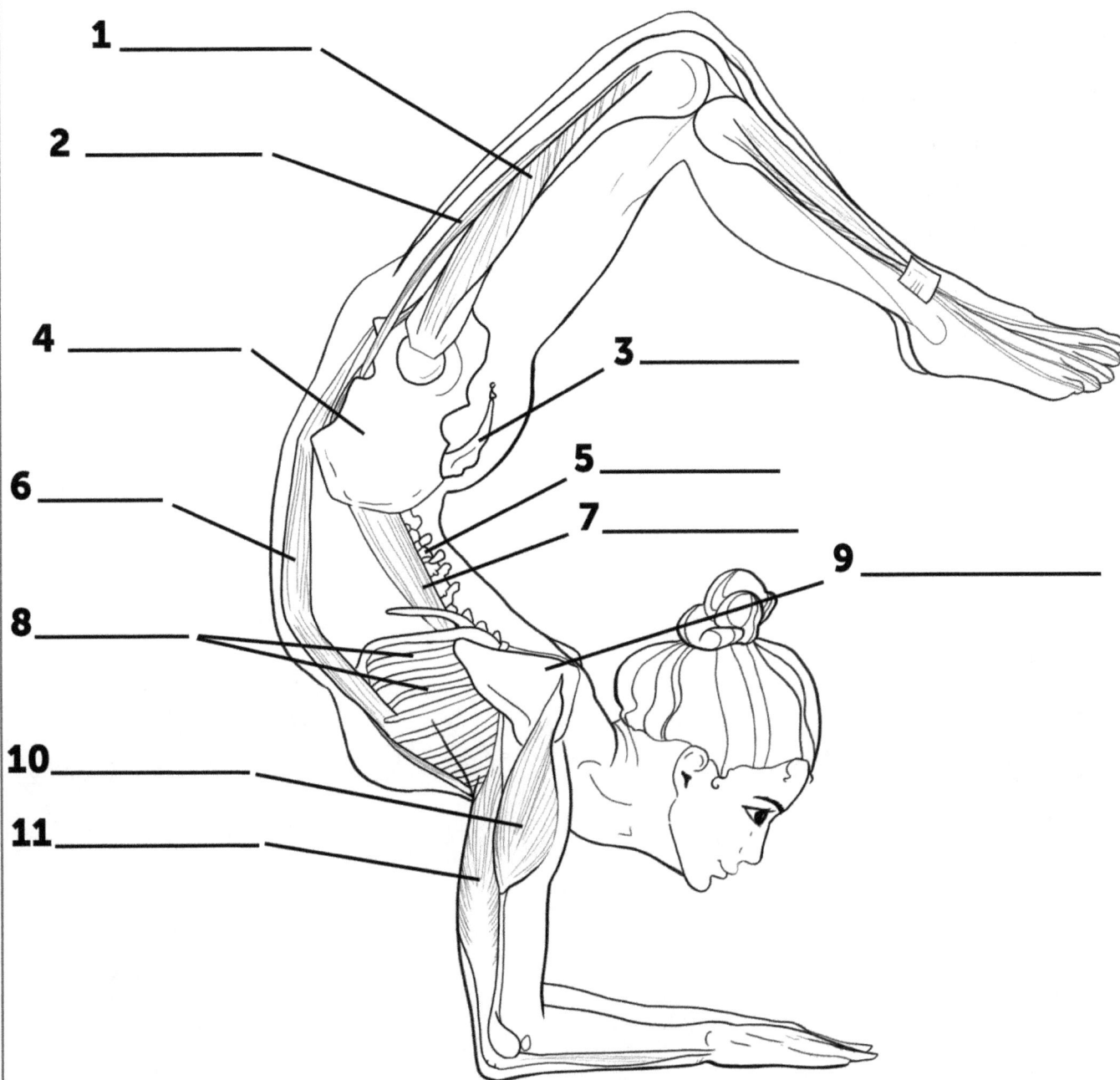

1 _____

2 _____

4 _____

3 _____

5 _____

7 _____

6 _____

9 _____

8 _____

10 _____

11 _____

110. POSTURA DEL ESCORPIÓN

1. MÚSCULO VASTO LATERAL
2. RECTO FEMORAL
3. HUESO SACRO
4. PELVIS
5. COLUMNA VERTEBRAL
6. RECTO ABDOMINAL
7. MÚSCULO PSOAS MAYOR
8. COSTILLAS
9. ESCÁPULA
10. DELTOIDES
11. TRÍCEPS BRAQUIAL

111. POSTURA DE LUCIÉRNAGA

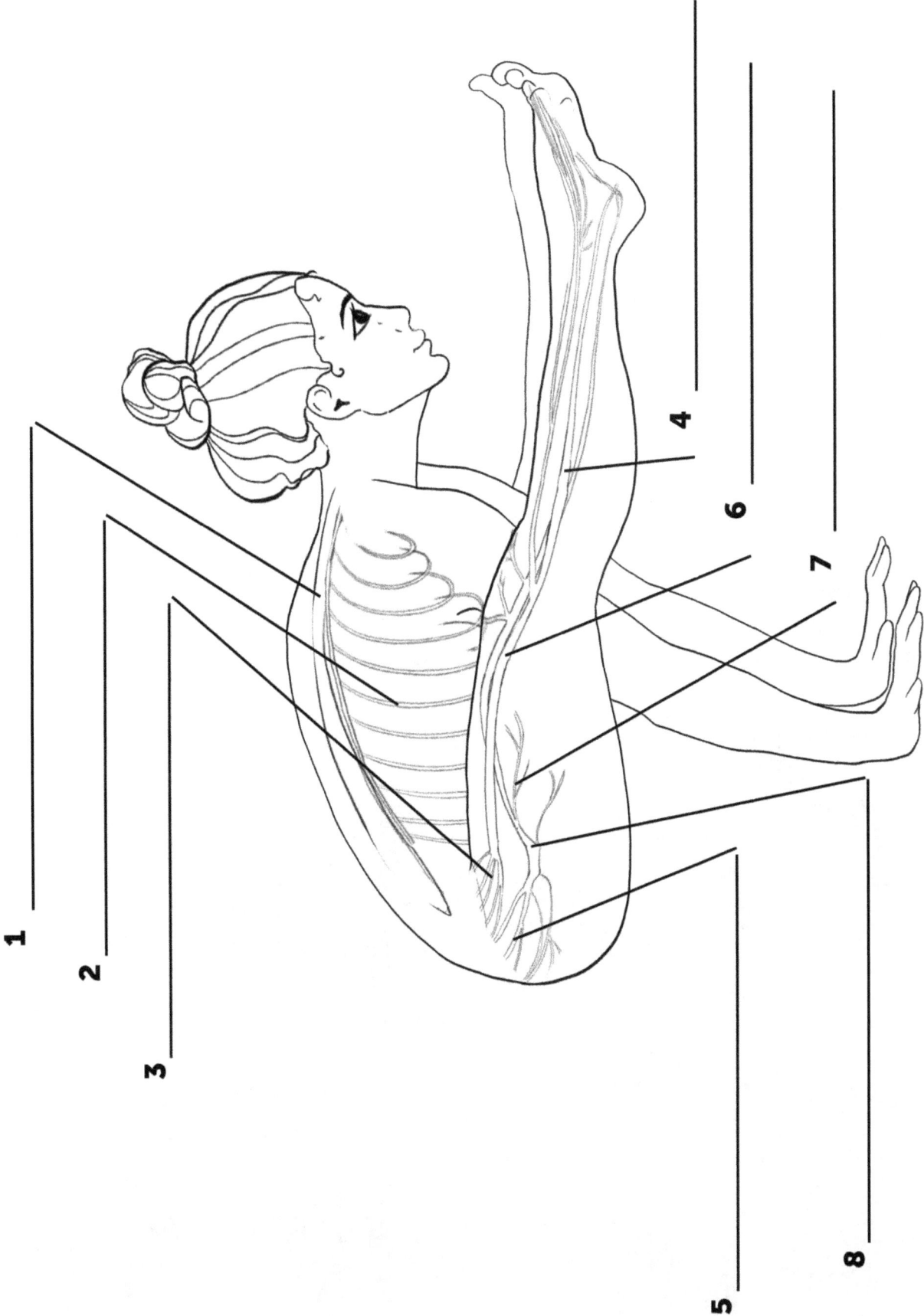

1

2

3

4

5

6

7

8

111. POSTURA DE LUCIÉRNAGA

1. MÉDULA ESPINAL

2. INTERCOSTALES

3. PLEXO SACRO

4. TIBIAL

5. PLEXO LUMBAR

6. CIÁTICO

7. RAMAS MUSCULARES DE FEMORAL

8. FEMORAL

112. POSTURA DE AVE DEL PARAÍSO

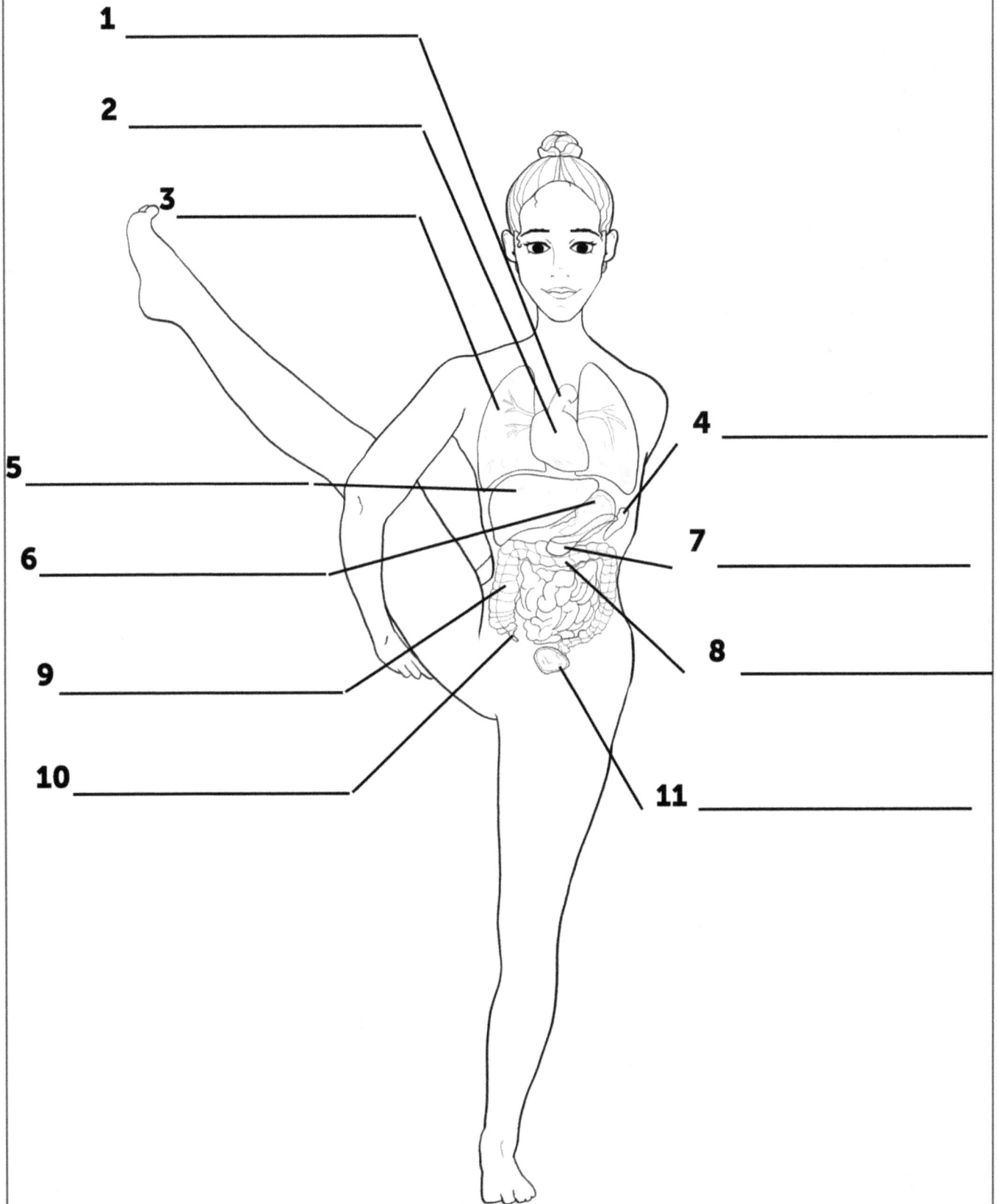

1 _____

2 _____

3 _____

4 _____

5 _____

6 _____

7 _____

8 _____

9 _____

10 _____

11 _____

112. POSTURA DE AVE DEL PARAÍSO

1. AORTA

2. CORAZÓN

3. PULMONES

4. BAZO

5. HÍGADO

6. ESTÓMAGO

7. PÁNCREAS

8. COLON TRANSVERSO

9. COLON ASCENDENTE

10. APÉNDICE

11. VEJIGA URINARIA

113. MAYURASANA

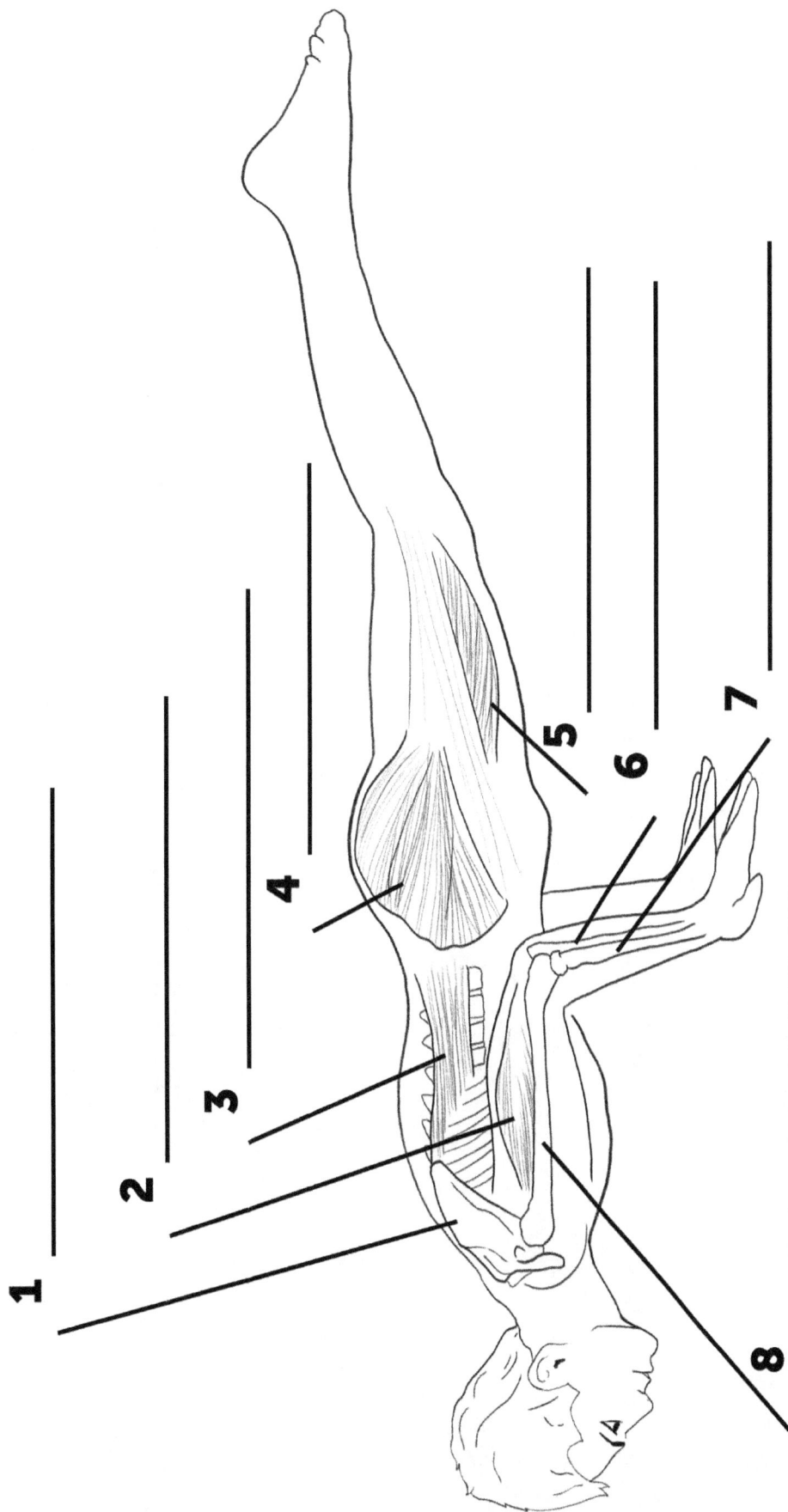

1
2
3
4
5
6
7
8

113. MAYURASANA

1. ESCÁPULA
2. TRÍCEPS BRAQUIAL
3. ERECTOR DE LA COLUMNA
4. MÚSCULO GLÚTEO MAYOR
5. CUADRÍCEPS
6. CÚBITO
7. RADIO
8. HÚMERO

114. POSTURA DE LA PALOMA REY CON UNA SOLA PIERNA II

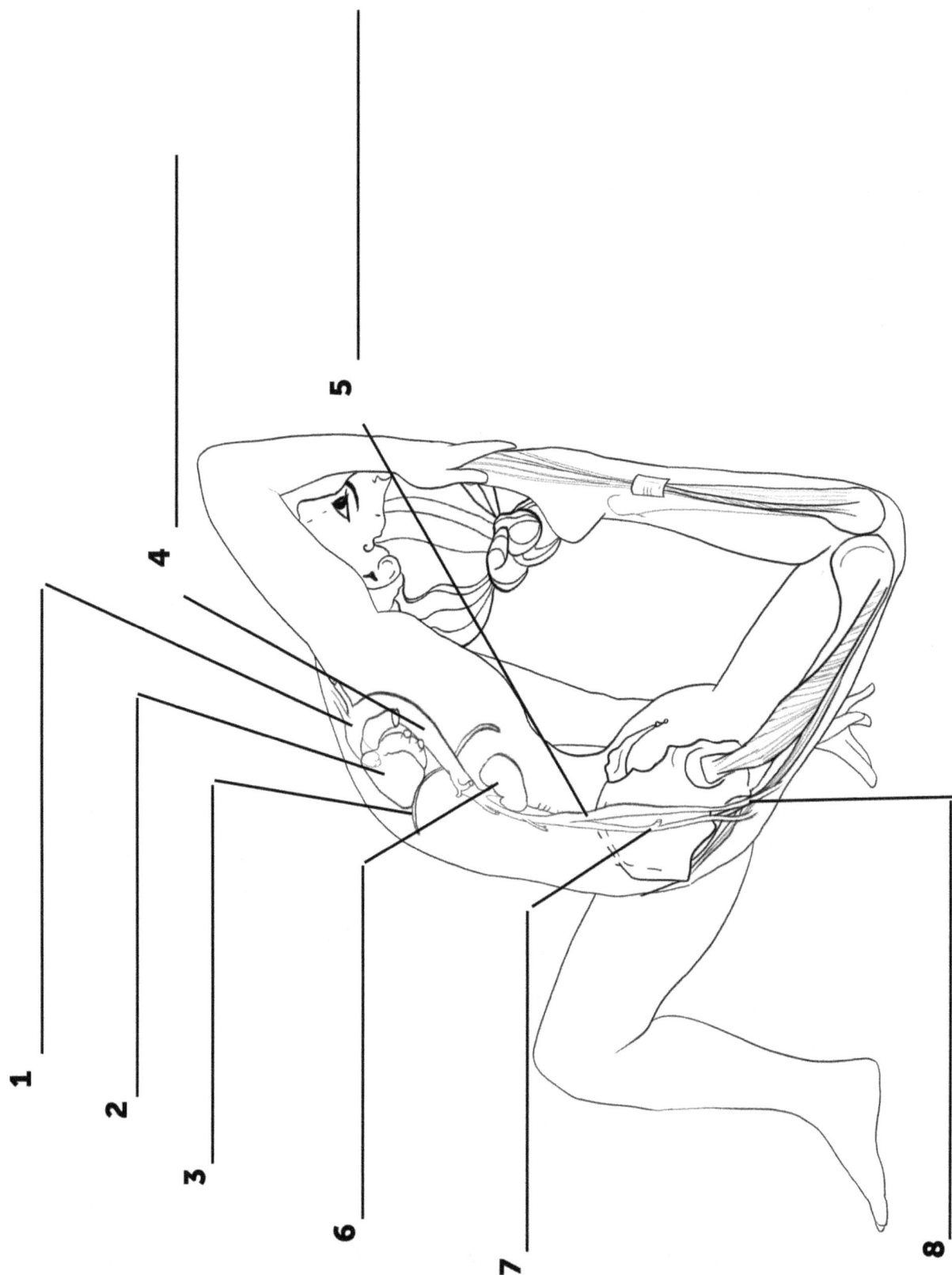

1

2

3

4

5

6

7

8

114. POSTURA DE LA PALOMA REY CON UNA SOLA PIERNA II

1.   AORTA TORÁCICA ASCENDENTE

2.   CORAZÓN

3.   DIAFRAGMA

4.   AORTA TORÁCICA DESCENDENTE

5.   AORTA ABDOMINAL

6.   RIÑÓN

7.   ARTERIA ILIACA COMÚN

8.   ARTERIA FEMORAL

115. LAGHU VAJRASANA

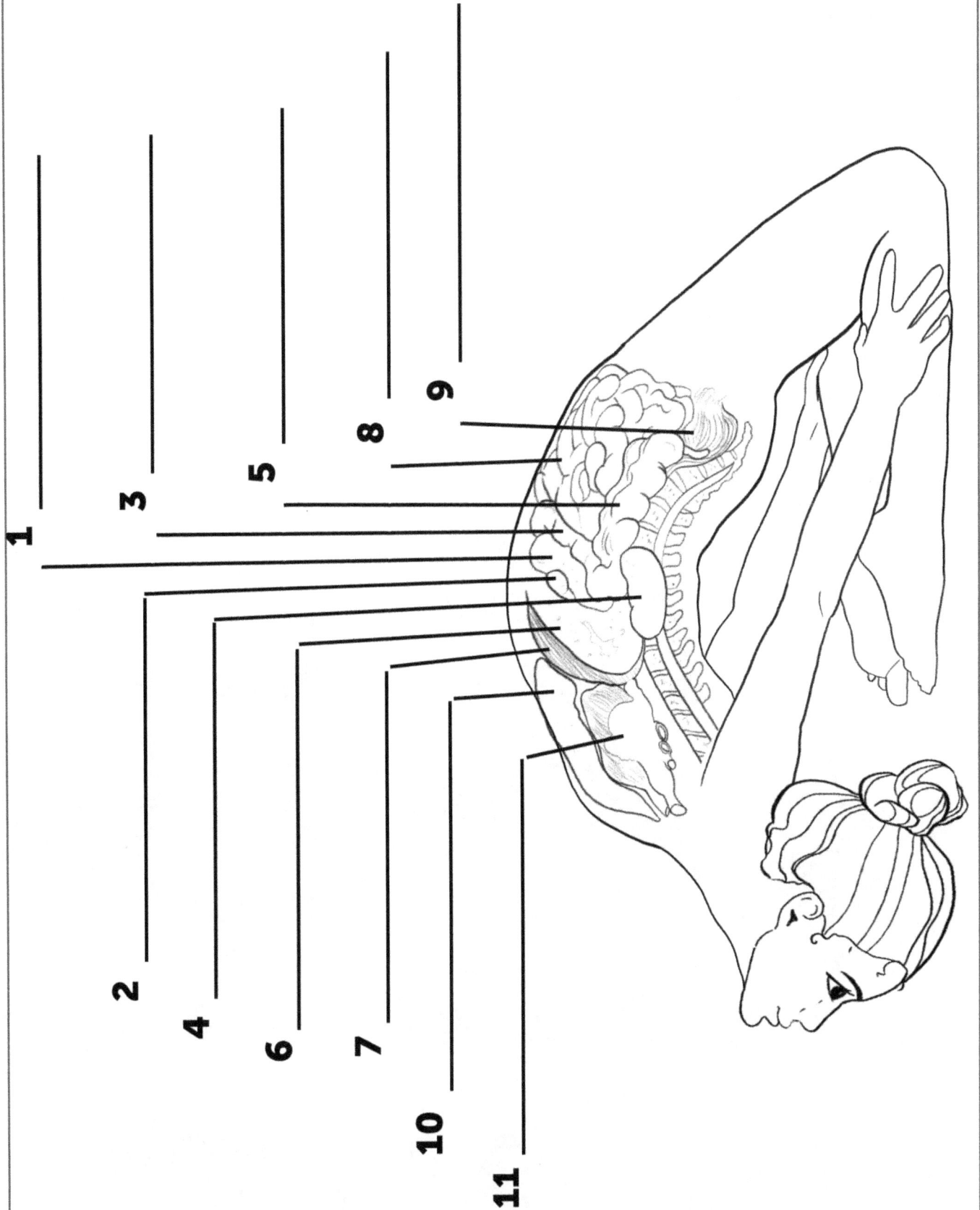

1

2

3

4

5

6

7

8

9

10

11

115. LAGHU VAJRASANA

1. ESTÓMAGO
2. VESÍCULA BILIAR
3. COLON TRANSVERSO
4. RIÑÓN
5. COLON ASCENDENTE
6. HÍGADO
7. DIAFRAGMA
8. FOLICULOS DE INTESTINO DELGADO
9. RECTO
10. PULMONES
11. CORAZÓN

116. PARIGHASANA

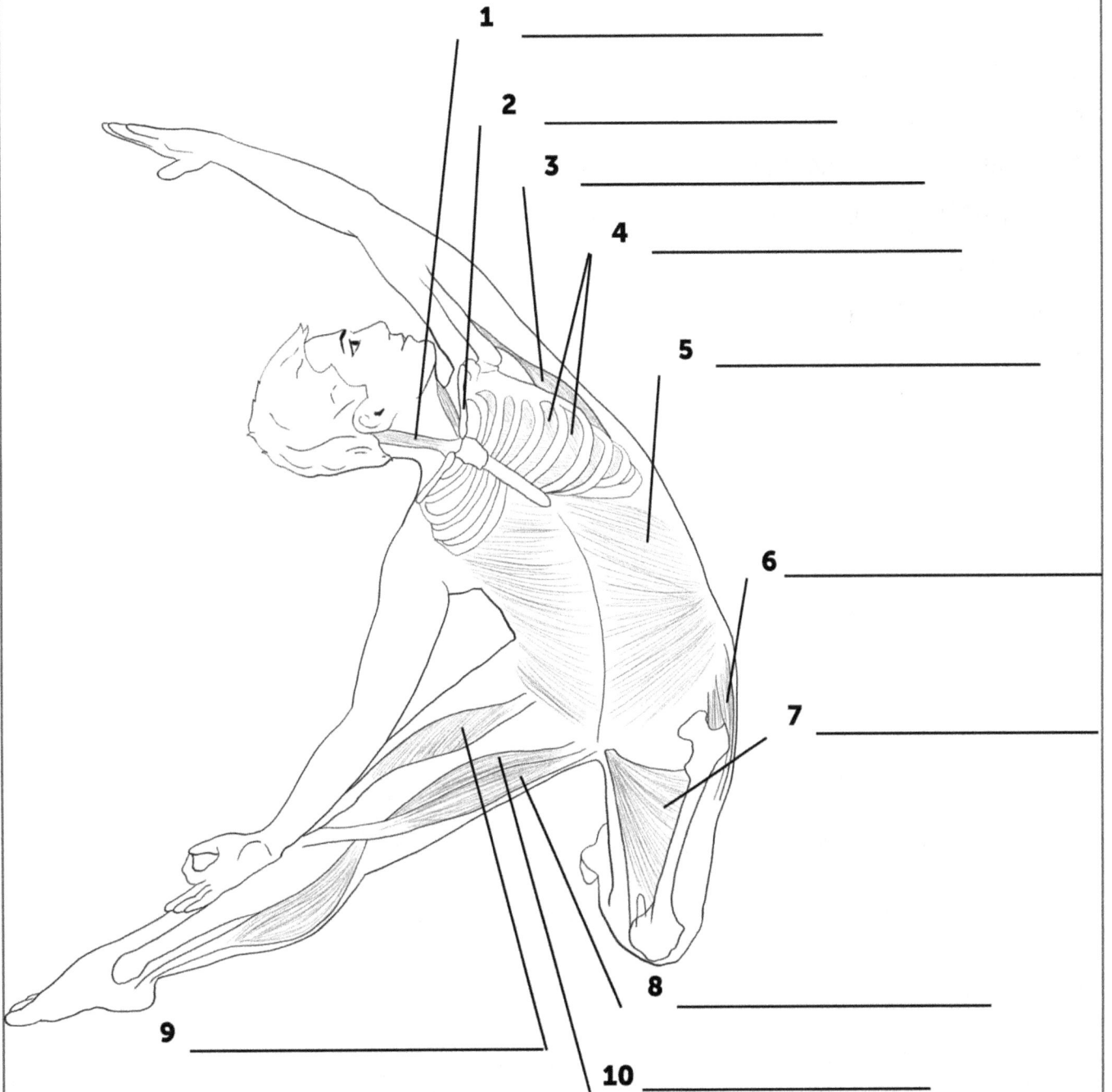

1 _____

2 _____

3 _____

4 _____

5 _____

6 _____

7 _____

8 _____

9 _____

10 _____

116. PARIGHASANA

1. MÚSCULO ESPLENIO DE LA CABEZA
2. CLAVÍCULA
3. LATISSIMUS DORSI
4. INTERCOSTALES
5. OBLICUO EXTERNO
6. MÚSCULO TENSOR DE LA FASCIA LATA
7. MÚSCULO ADUCTOR LARGO DEL MUSLO
8. GRÁCIL
9. RECTO FEMORAL
10. ADUCTOR MAYOR

117. POSTURA SAGE KOUNDIYA I

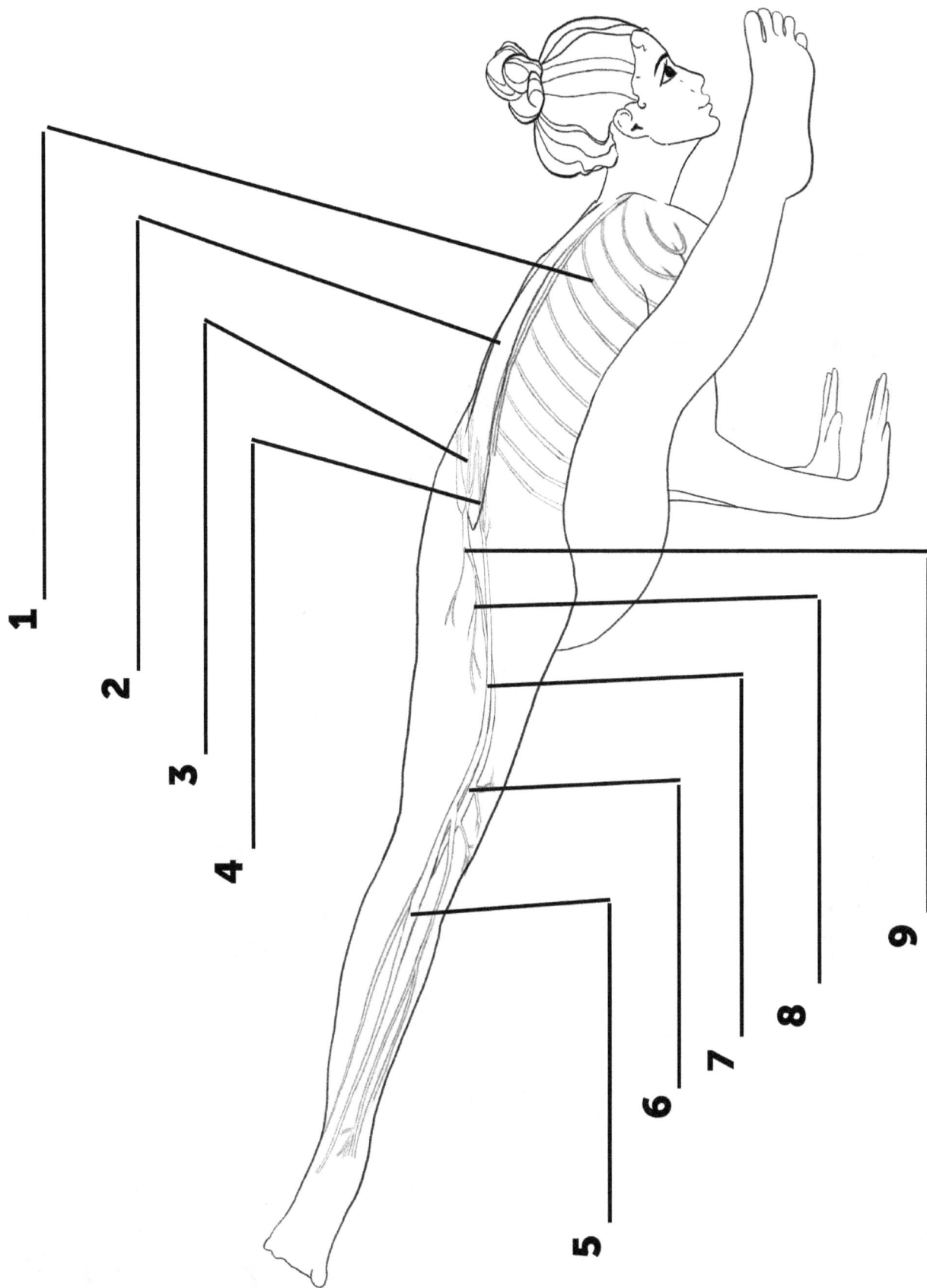

1

2

3

4

5

6

7

8

9

117. POSTURA SAGE KOUNDIYA I

1. INTERCOSTALES
2. MÉDULA ESPINAL
3. PLEXO LUMBAR
4. PLEXO SACRO
5. TIBIAL
6. SAFENA
7. CIÁTICO
8. RAMAS MUSCULARES DE FEMORAL
9. FEMORAL

118. POSTURA SAGE KOUNDIYA II

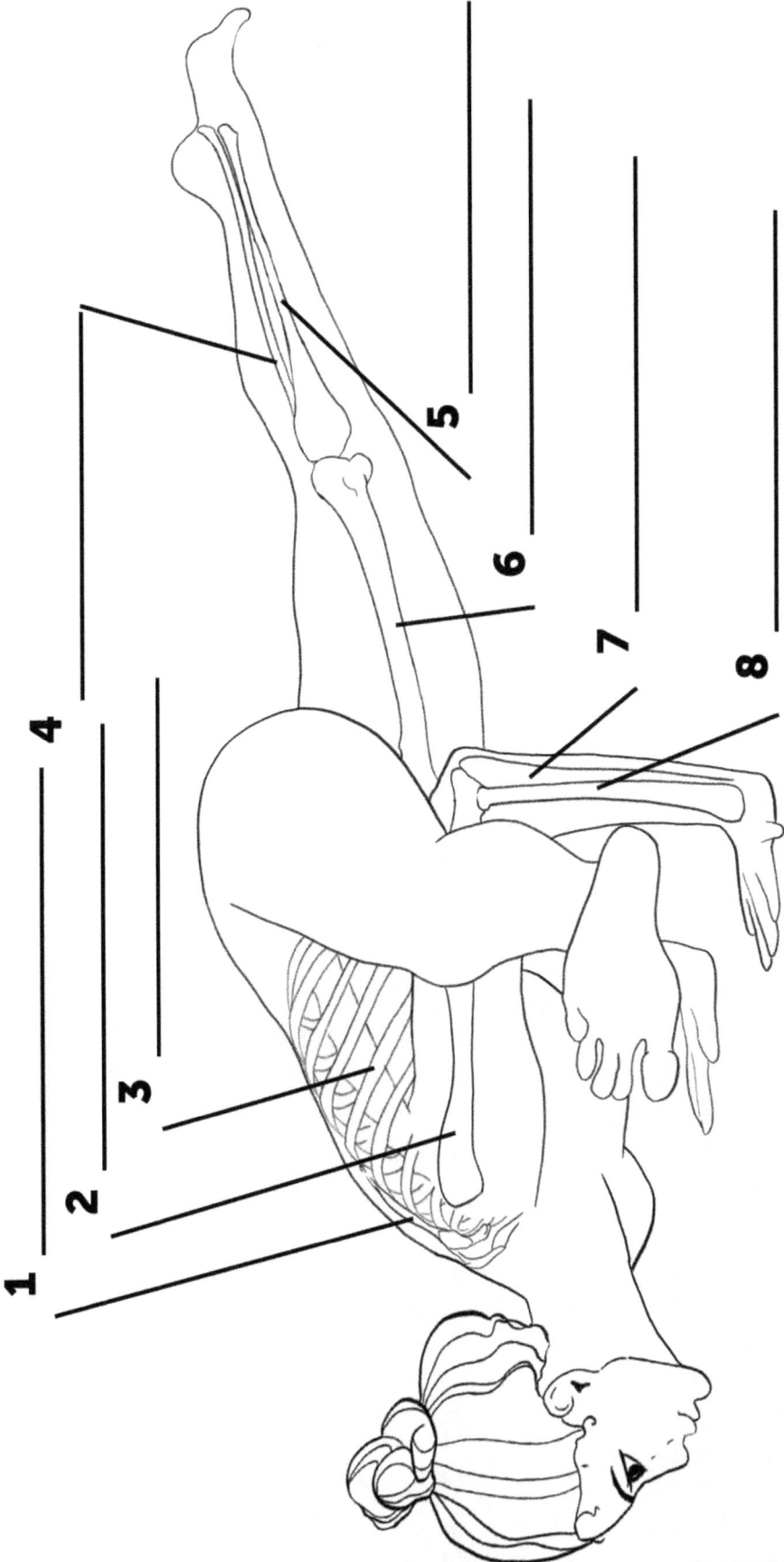

118. POSTURA SAGE KOUNDIYA II

1. ESCÁPULA
2. HÚMERO
3. COSTILLAS
4. PERONÉ
5. TIBIA
6. FÉMUR
7. CÚBITO
8. RADIO

119. EKA PADA SIRSASANA

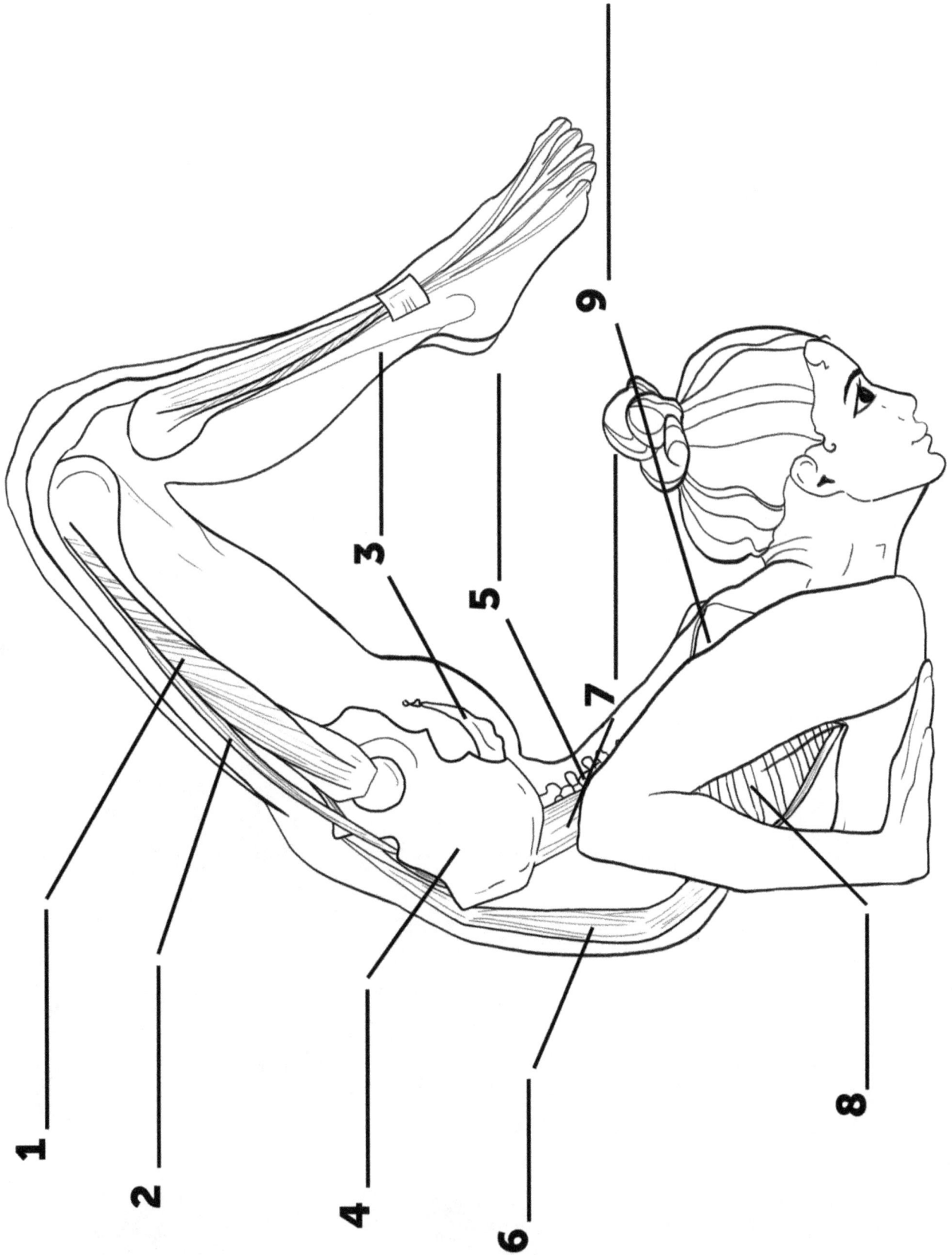

119. EKA PADA SIRSASANA

1. MÚSCULO VASTO LATERAL

2. RECTO FEMORAL

3. SACRO

4. PELVIS

5. COLUMNA VERTEBRAL

6. RECTO ABDOMINAL

7. ERECTOR DE LA COLUMNA

8. COSTILLAS

9. ESCÁPULA

120. POSTURA DEL SALTAMONTES BEBÉ

MAESTRO

1
2
3
4
5
6
7
8

120. POSTURA DEL SALTAMONTES BEBÉ MAESTRO

1. CUADRÍCEPS
2. TRÍCEPS BRAQUIAL
3. BÍCEPS BRAQUIAL
4. TRAPECIO
5. DELTOIDES
6. TIBIAL ANTERIOR
7. GASTROCNEMIO
8. PRONADORES

121. DWI PADA VIPARITA DANDASANA

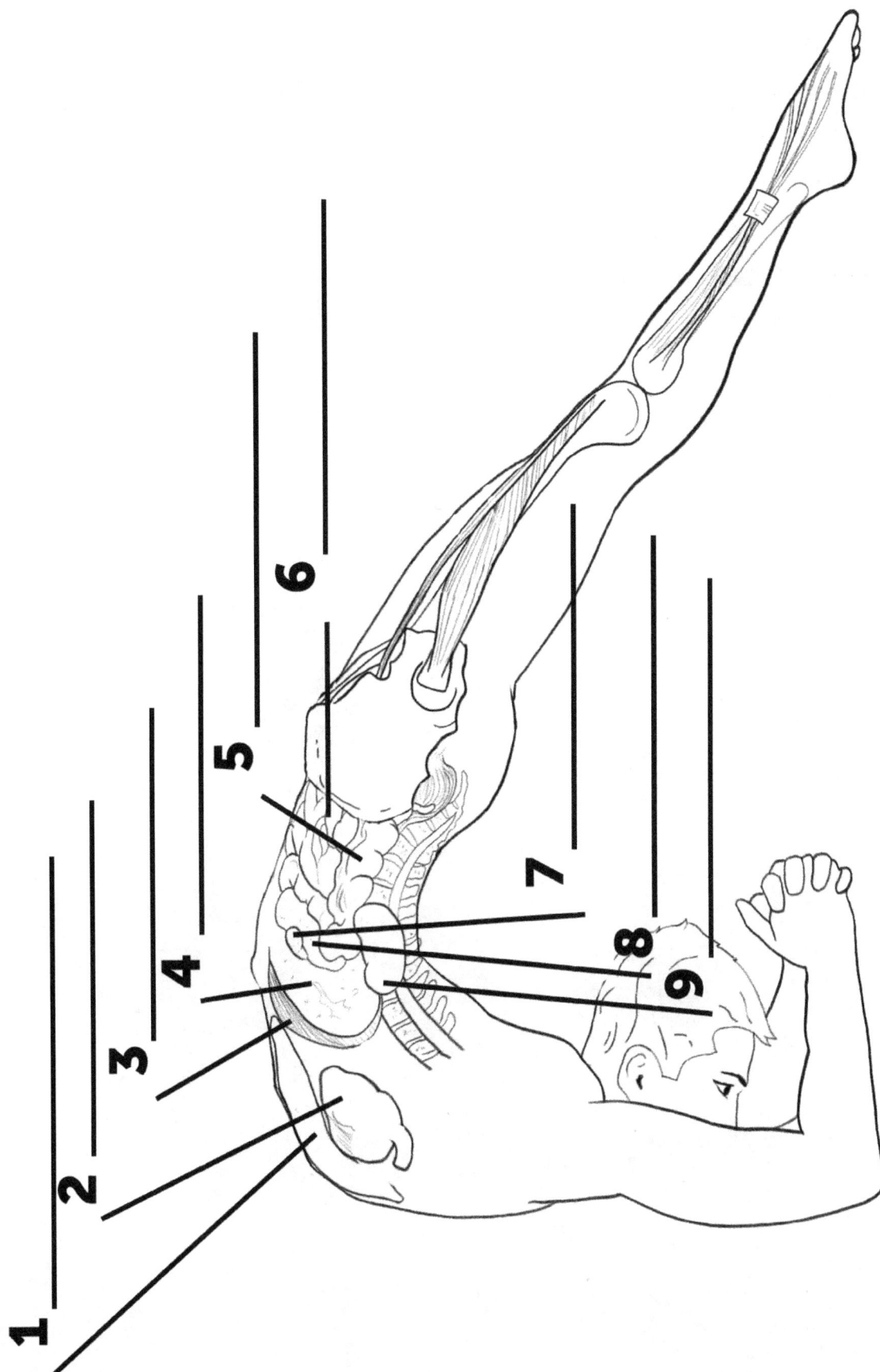

121. DWI PADA VIPARITA DANDASANA

1. PULMONES
2. CORAZÓN
3. DIAFRAGMA
4. HÍGADO
5. COLON ASCENDENTE
6. FOLICULOS DE INTESTINO DELGADO
7. VESÍCULA BILIAR
8. ESTÓMAGO
9. RIÑÓN

122. TORSIÓN BHARADVAJA

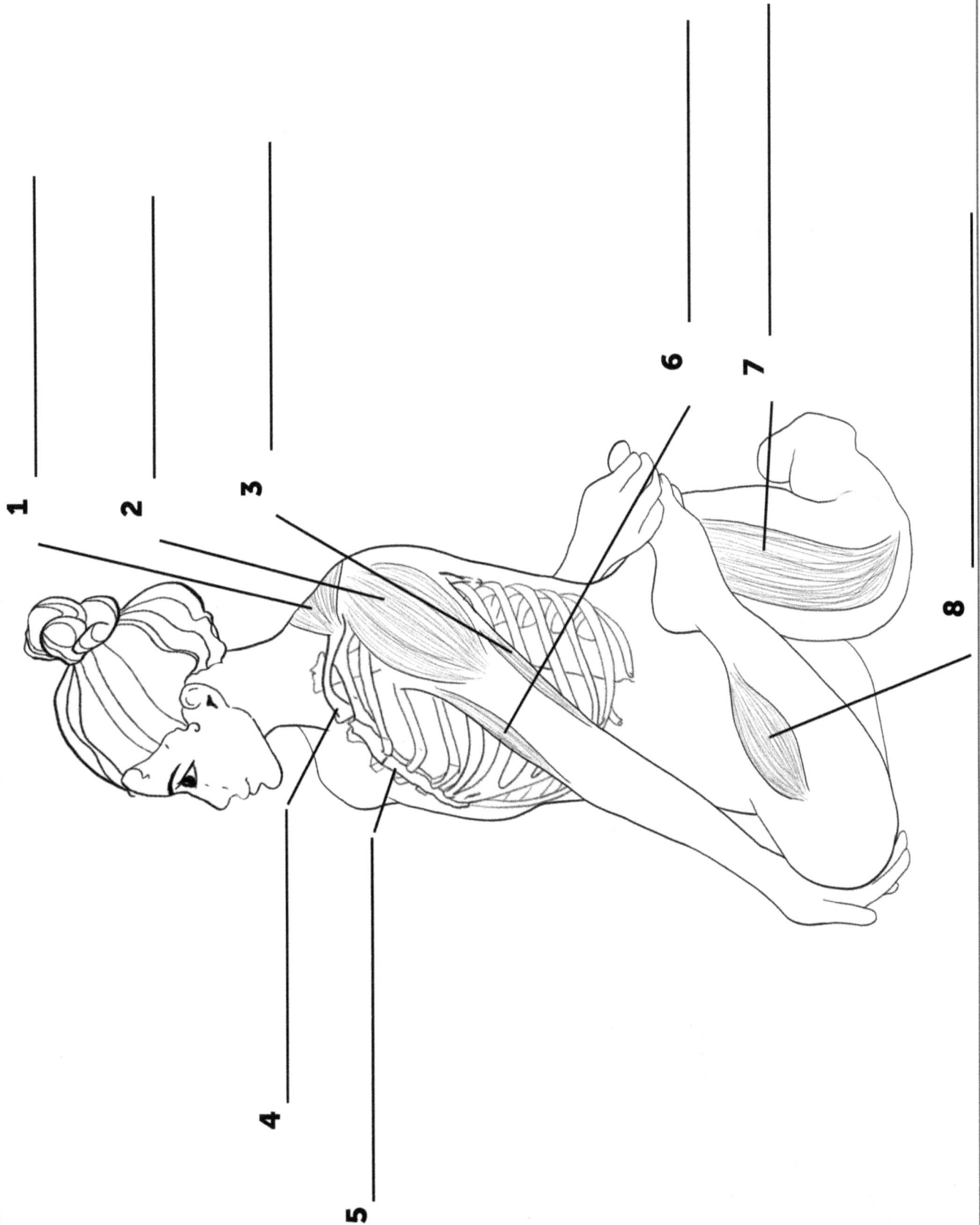

1

2

3

4

5

6

7

8

122. TORSIÓN BHARADVAJA

1.   TRAPECIO

2.   DELTOIDES

3.   TRÍCEPS BRAQUIAL

4.   CLAVÍCULA

5.   ESTERNÓN

6.   BÍCEPS BRAQUIAL

7.   CUADRÍCEPS

8.   GASTROCNEMIO

123. ASTAVAKRASANA

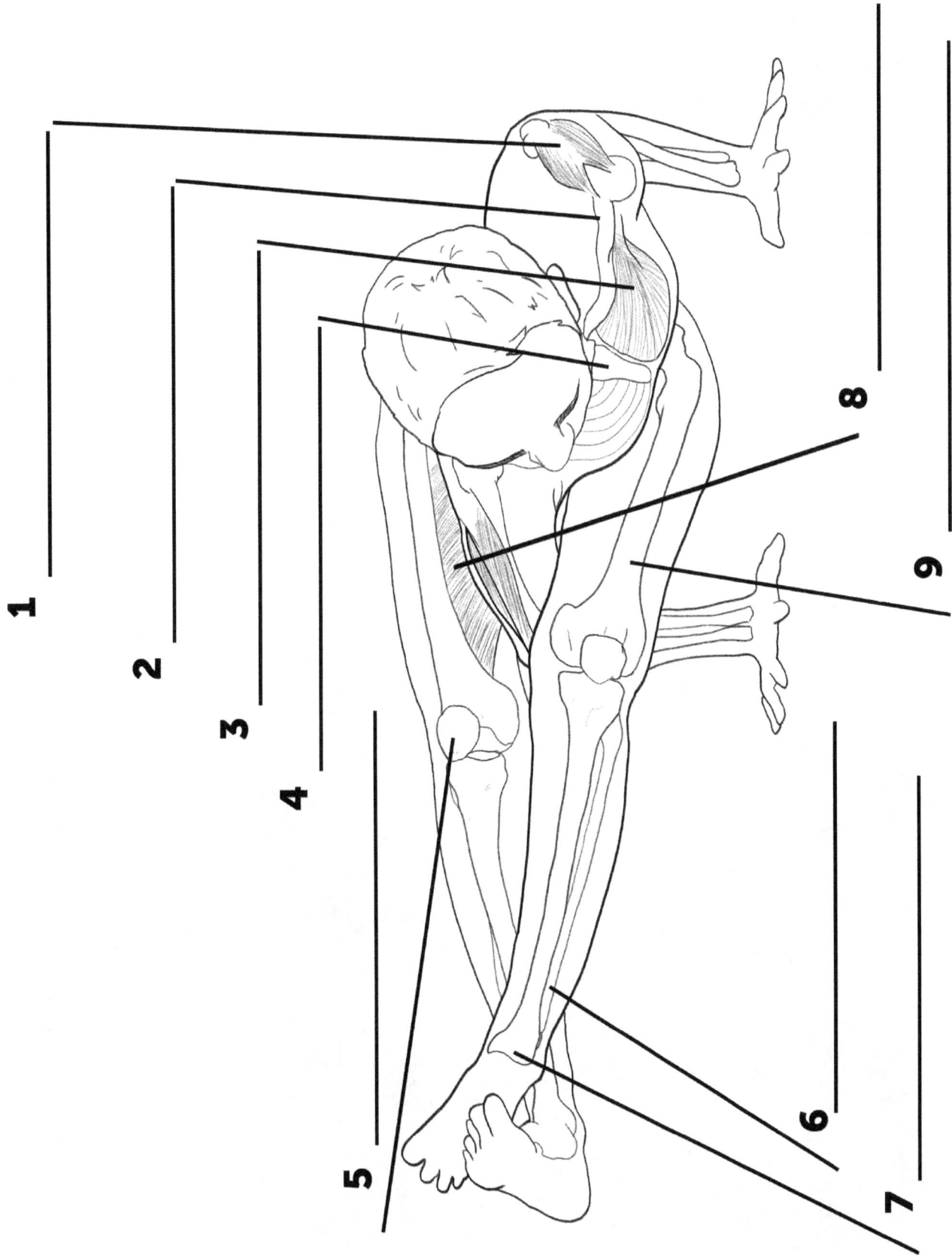

1

2

3

4

5

6

7

8

9

123. ASTAVAKRASANA

1. TRÍCEPS BRAQUIAL
2. CLAVÍCULA
3. PECTORAL MAYOR
4. ESTERNÓN
5. RÓTULA
6. PERONÉ
7. TIBIA
8. ADUCTORES
9. FÉMUR

124. POSTURA DEL LOTO MEDIO ATADO DEL SABIO

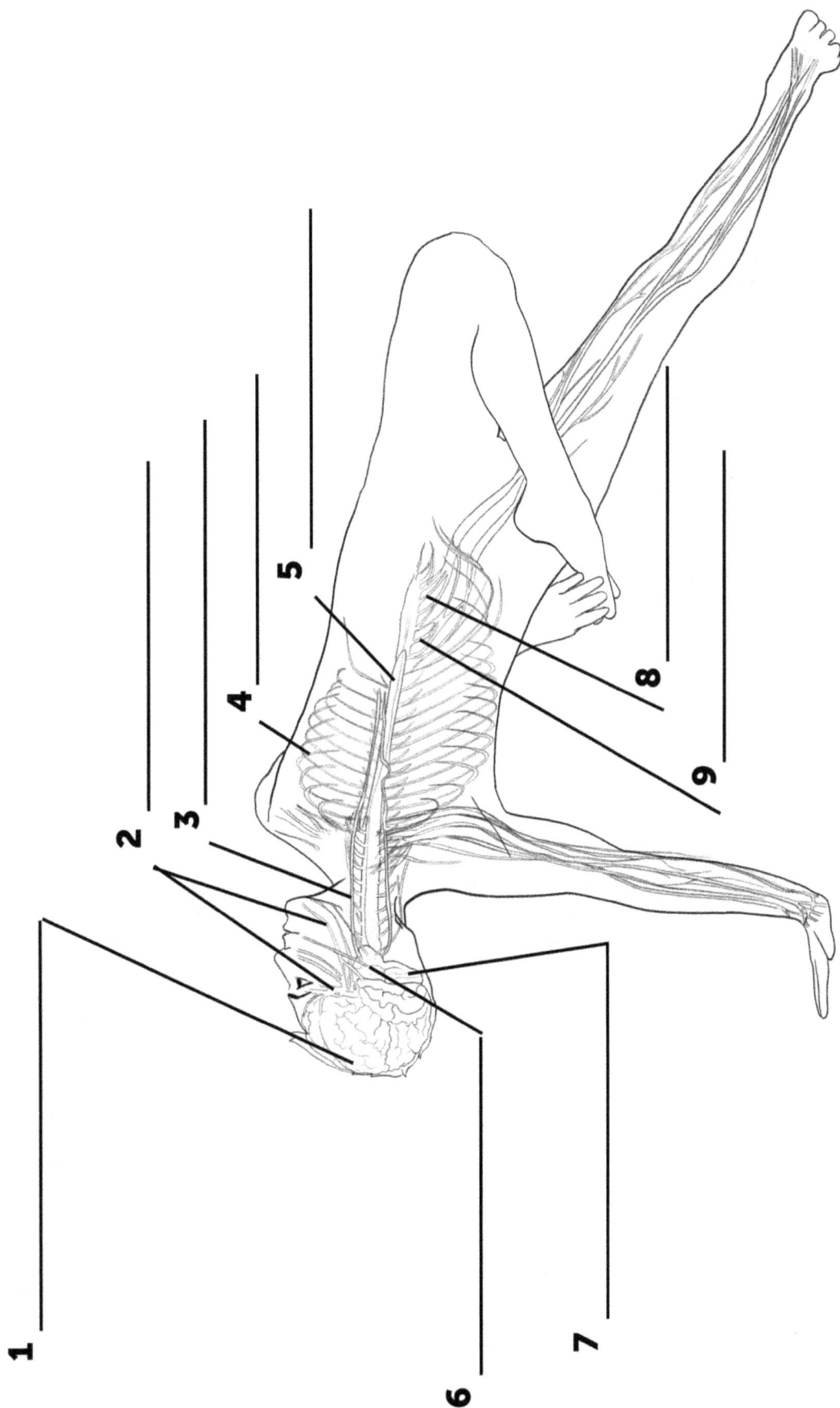

124. POSTURA DEL LOTO MEDIO ATADO DEL SABIO

1. CEREBRO
2. NERVIOS CRANEALES
3. NERVIO VAGO
4. INTERCOSTALES
5. MÉDULA ESPINAL
6. TRONCO ENCEFÁLICO
7. CEREBELO
8. PLEXO SACRO
9. PLEXO LUMBAR

125. BHUJAPIDASANA

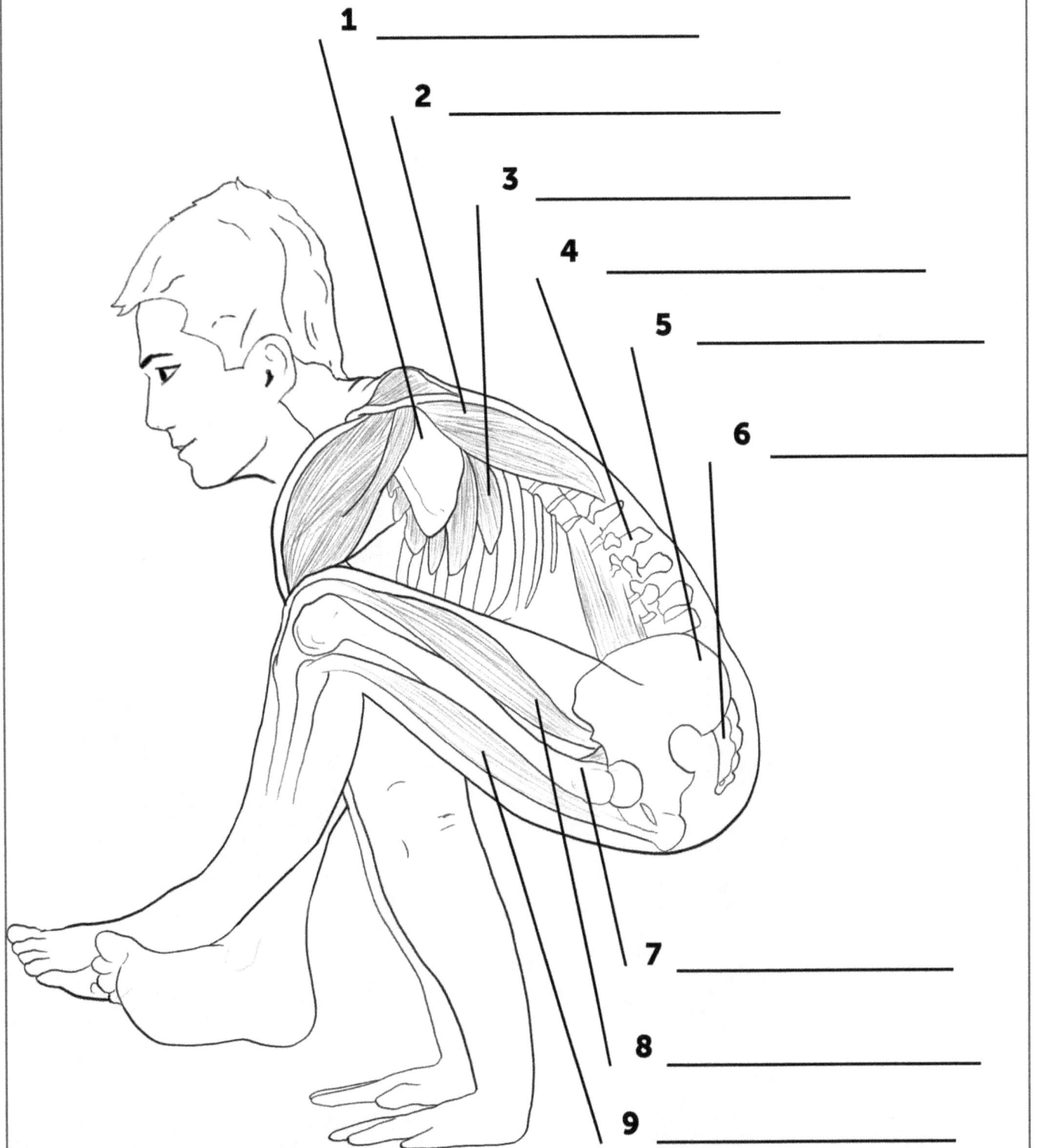

1 _____

2 _____

3 _____

4 _____

5 _____

6 _____

7 _____

8 _____

9 _____

125. BHUJAPIDASANA

1.	ESCÁPULA

2.	ROMBOIDES

3.	SERRATO ANTERIOR

4.	COLUMNA VERTEBRAL

5.	PELVIS

6.	SACRO

7.	FÉMUR

8.	CUADRÍCEPS

9.	ISQUIOTIBIALES

126. SUPER SOLDADO

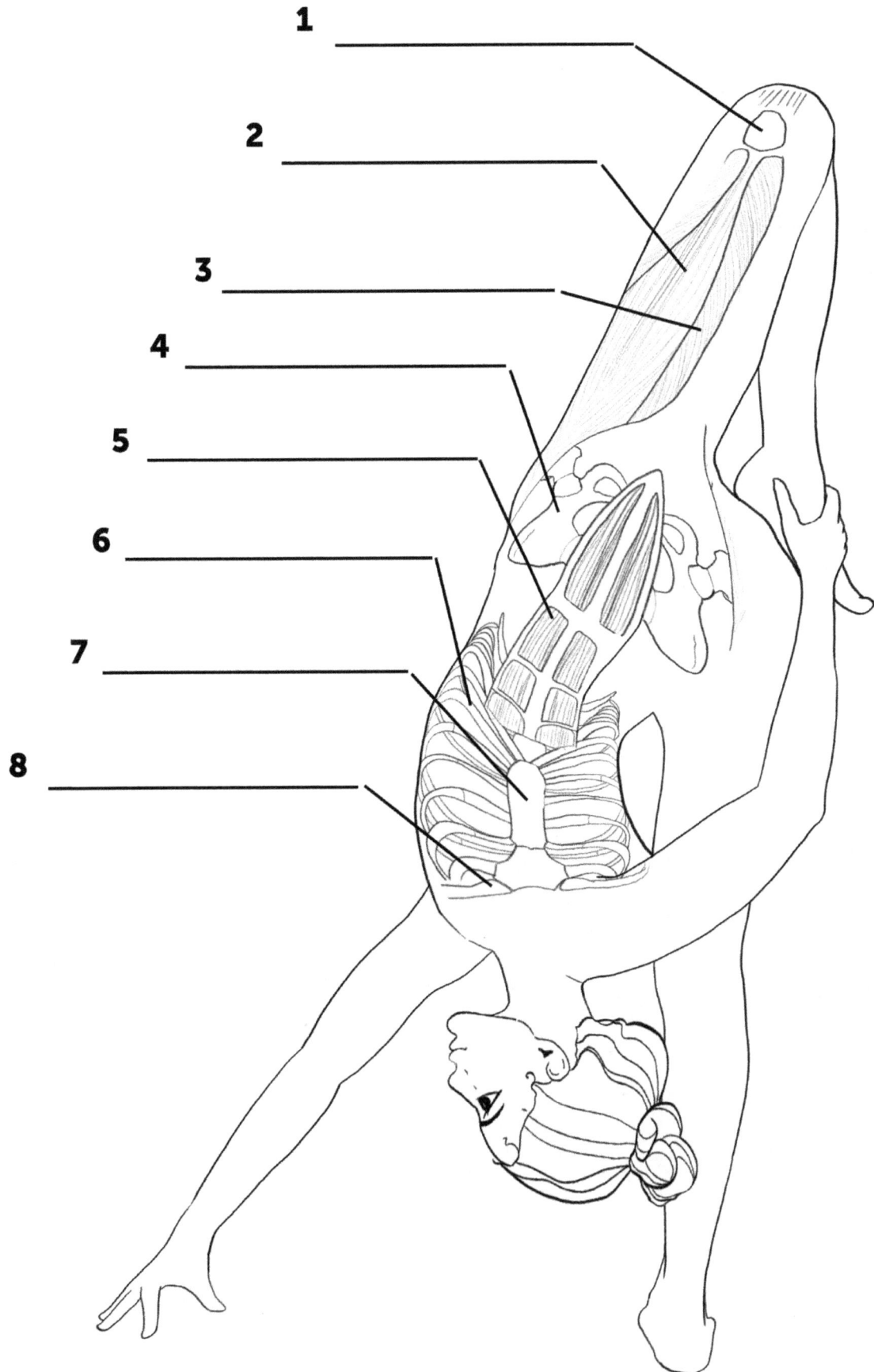

1 _____

2 _____

3 _____

4 _____

5 _____

6 _____

7 _____

8 _____

126. SUPER SOLDADO

1.   RÓTULA

2.   RECTO FEMORAL

3.   VASTO MEDIAL

4.   PELVIS

5.   RECTO ABDOMINAL

6.   COSTILLAS

7.   ESTERNÓN

8.   CLAVÍCULA

127. POSTURA DEL MONO

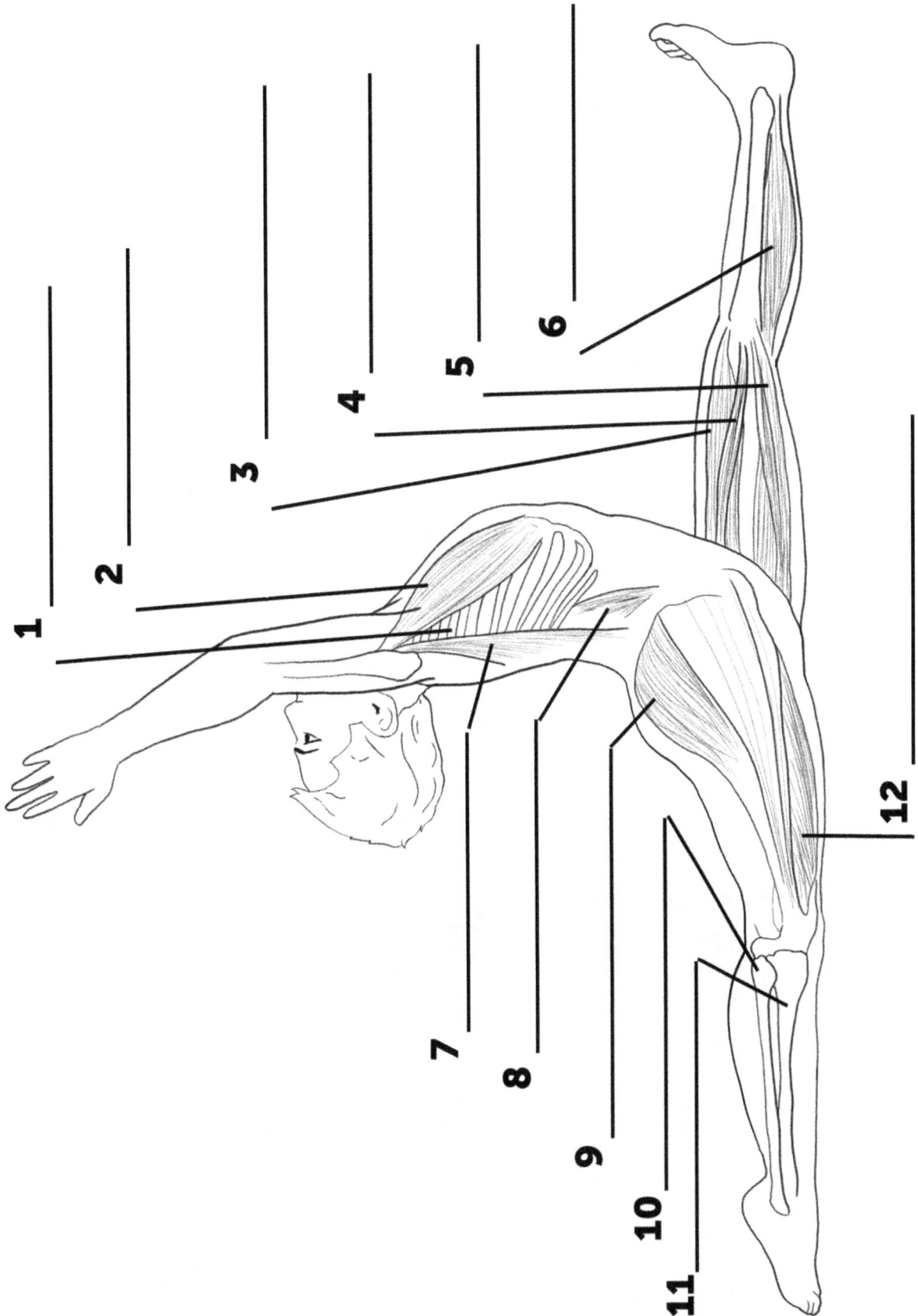

1

2

3

4

5

6

7

8

9

10

11

12

127. POSTURA DEL MONO

1. COSTILLAS
2. PECTORAL MAYOR
3. RECTO FEMORAL
4. SARTORIO
5. ISQUIOTIBIALES
6. GASTROCNEMIO
7. LATISSIMUS DORSI
8. ERECTOR DE LA COLUMNA
9. MÚSCULO GLÚTEO MAYOR
10. PERONÉ
11. TIBIA
12. CUADRÍCEPS

128. EUPAVISTHA KONASAN

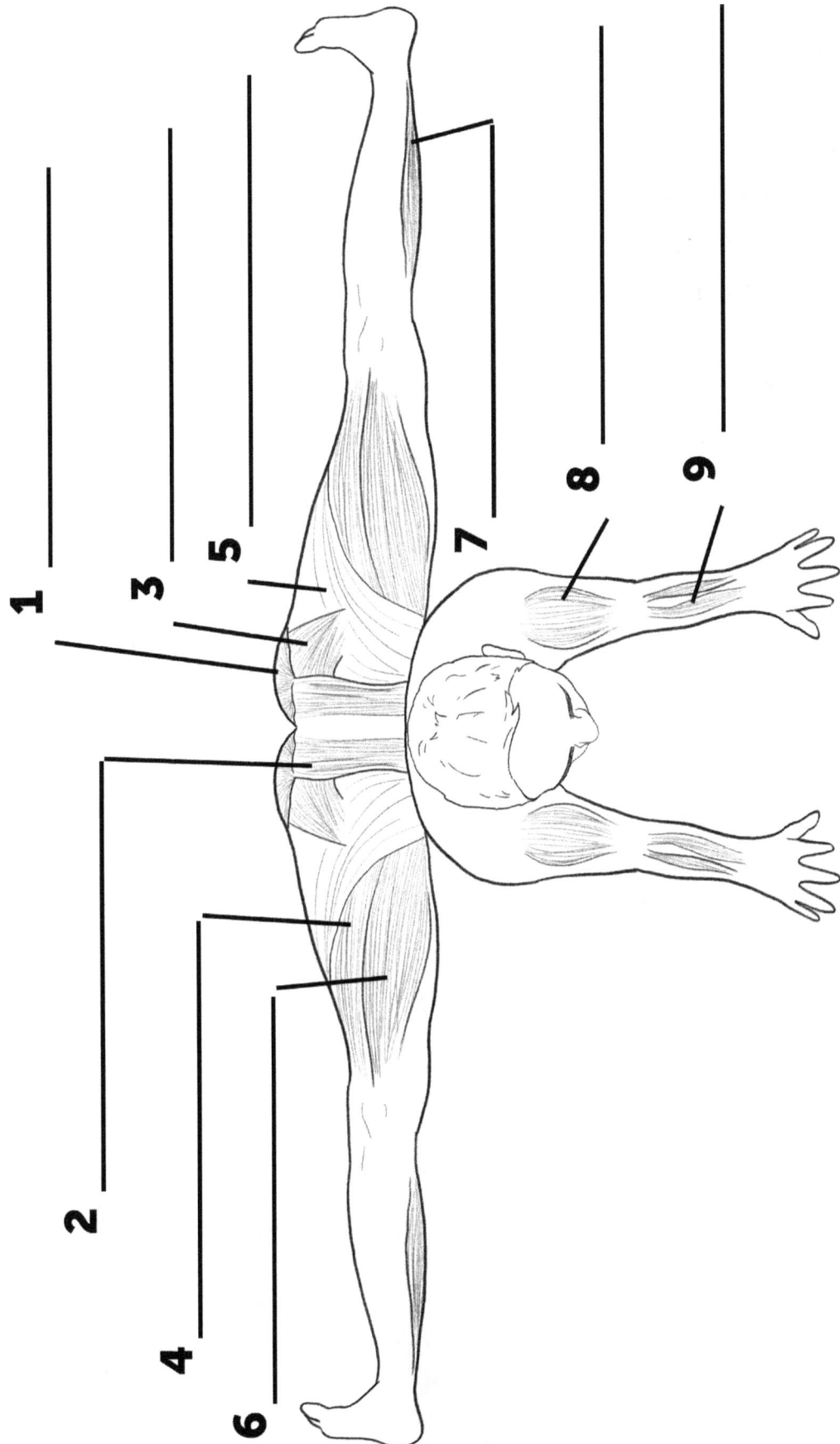

1

2

3

4

5

6

7

8

9

128. EUPAVISTHA KONASAN

1. MÚSCULO GLÚTEO MAYOR
2. ERECTOR DE LA COLUMNA
3. GLÚTEO MEDIO
4. MÚSCULO VASTO LATERAL
5. BANDA ILIOTIBIAL
6. RECTO FEMORAL
7. GASTROCNEMIO
8. DELTOIDES
9. PRONADORES

129. LAGARTO DE EQUILIBRIO EXTENDIDO

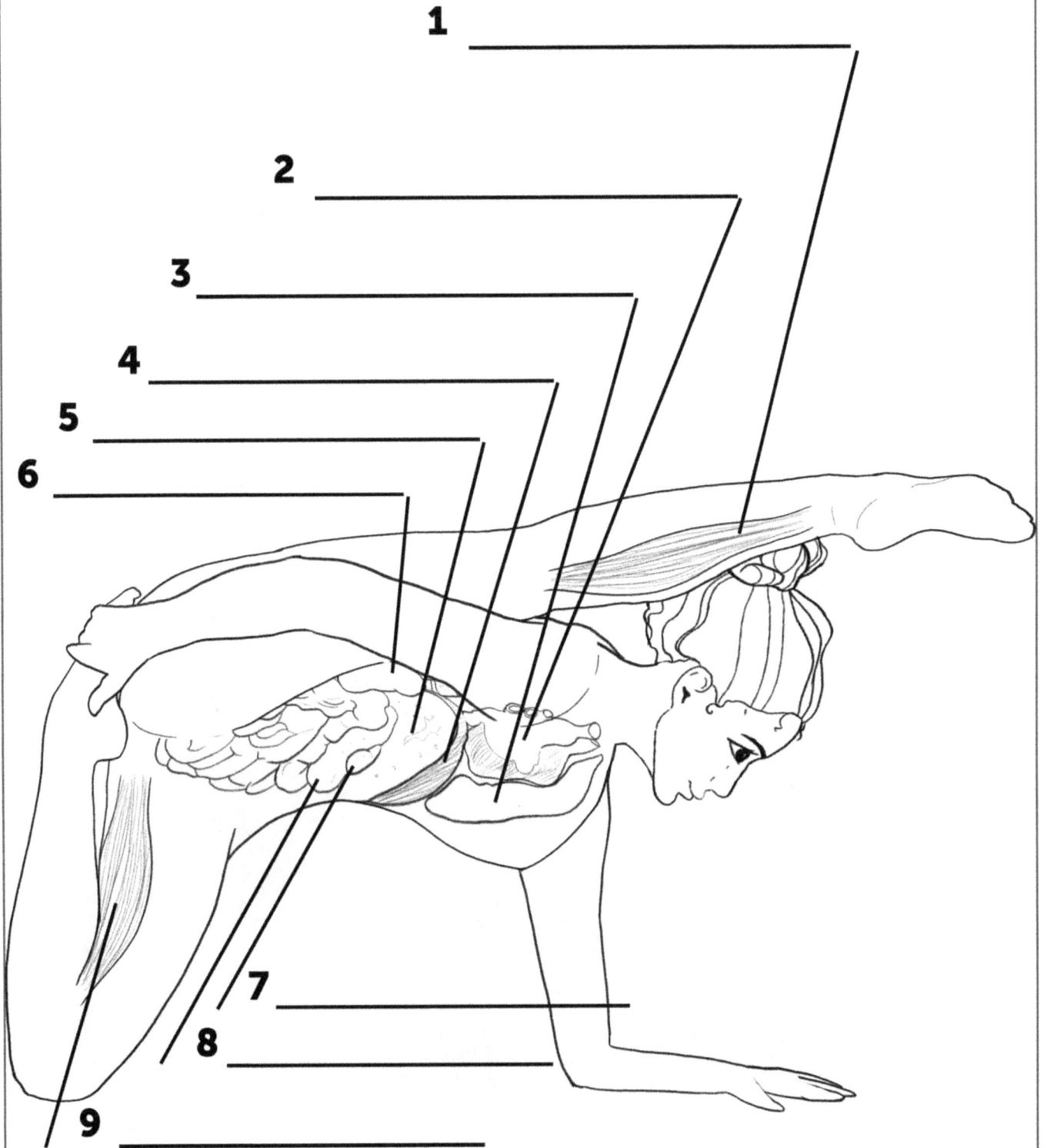

1 _____

2 _____

3 _____

4 _____

5 _____

6 _____

7 _____

8 _____

9 _____

129. LAGARTO DE EQUILIBRIO EXTENDIDO

1. GASTROCNEMIO
2. CORAZÓN
3. PULMONES
4. DIAFRAGMA
5. HÍGADO
6. RIÑÓN
7. VESÍCULA BILIAR
8. ESTÓMAGO
9. ISQUIOTIBIALES

130 KURMASANA

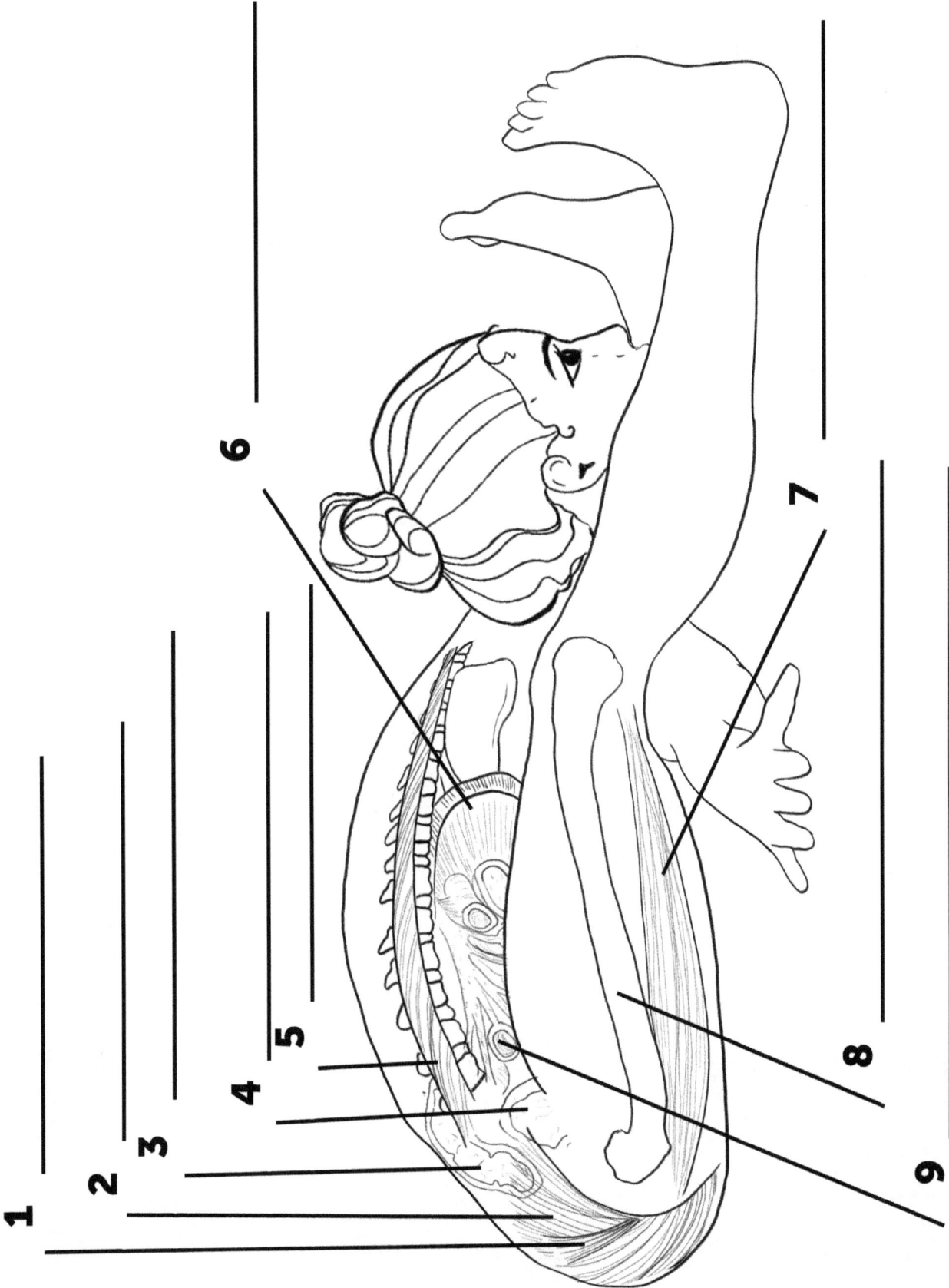

130 KURMASANA

1. PIRIFORME
2. MÚSCULO GLÚTEO MAYOR
3. RECTO
4. VEJIGA URINARIA
5. MÚSCULOS ESPINALES
6. DIAFRAGMA
7. ISQUIOTIBIALES
8. FÉMUR
9. FOLICULOS DE INTESTINO DELGADO

131. VIPARITA SALABHASANA

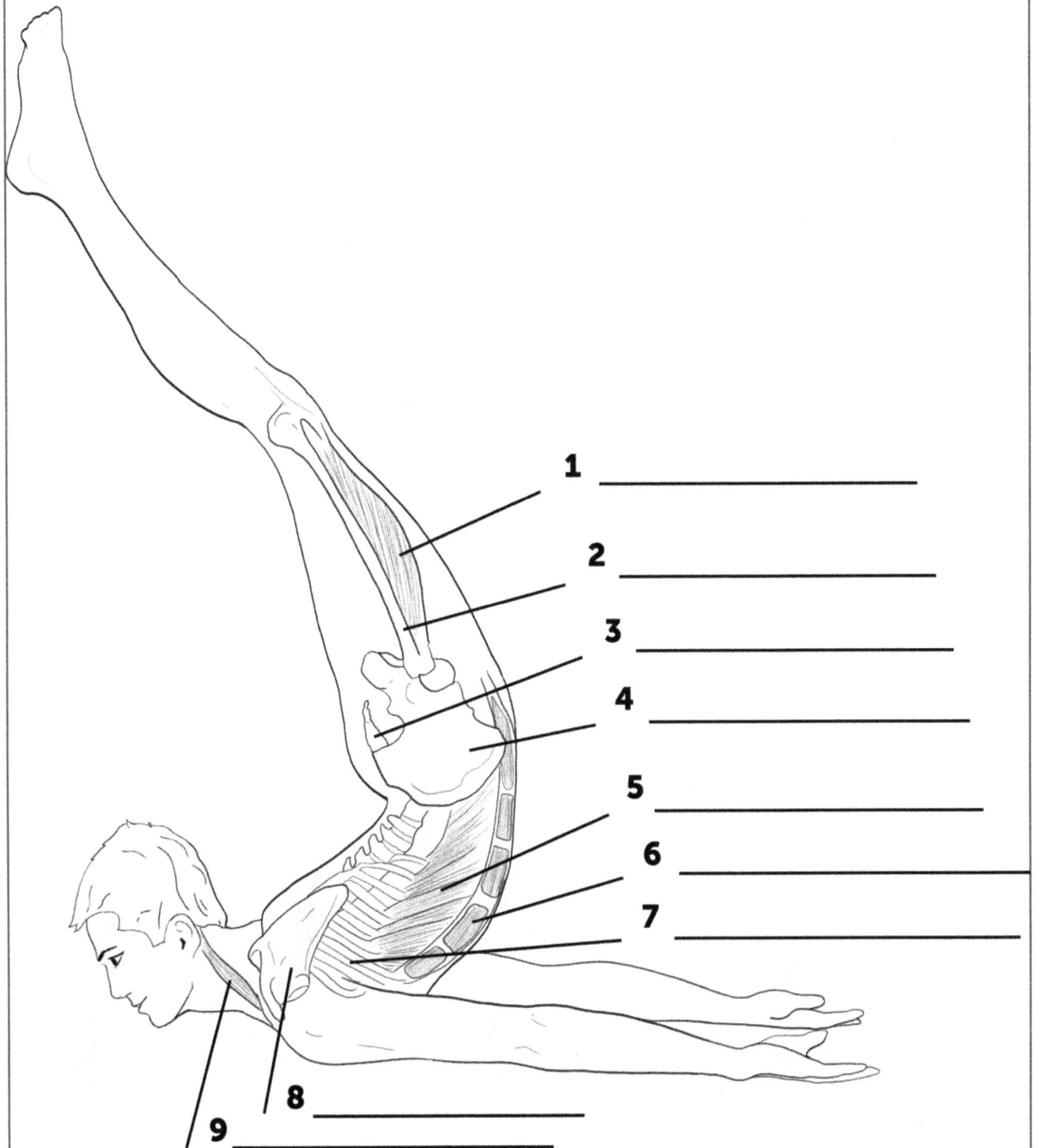

1 _____

2 _____

3 _____

4 _____

5 _____

6 _____

7 _____

8 _____

9 _____

131. VIPARITA SALABHASANA

1. CUADRÍCEPS
2. FÉMUR
3. SACRO
4. PELVIS
5. OBLICUO EXTERNO
6. RECTO ABDOMINAL
7. COSTILLAS
8. ESCÁPULA
9. ESTERNOCLEIDOMASTOIDEO

132. YOGANIDRASANA

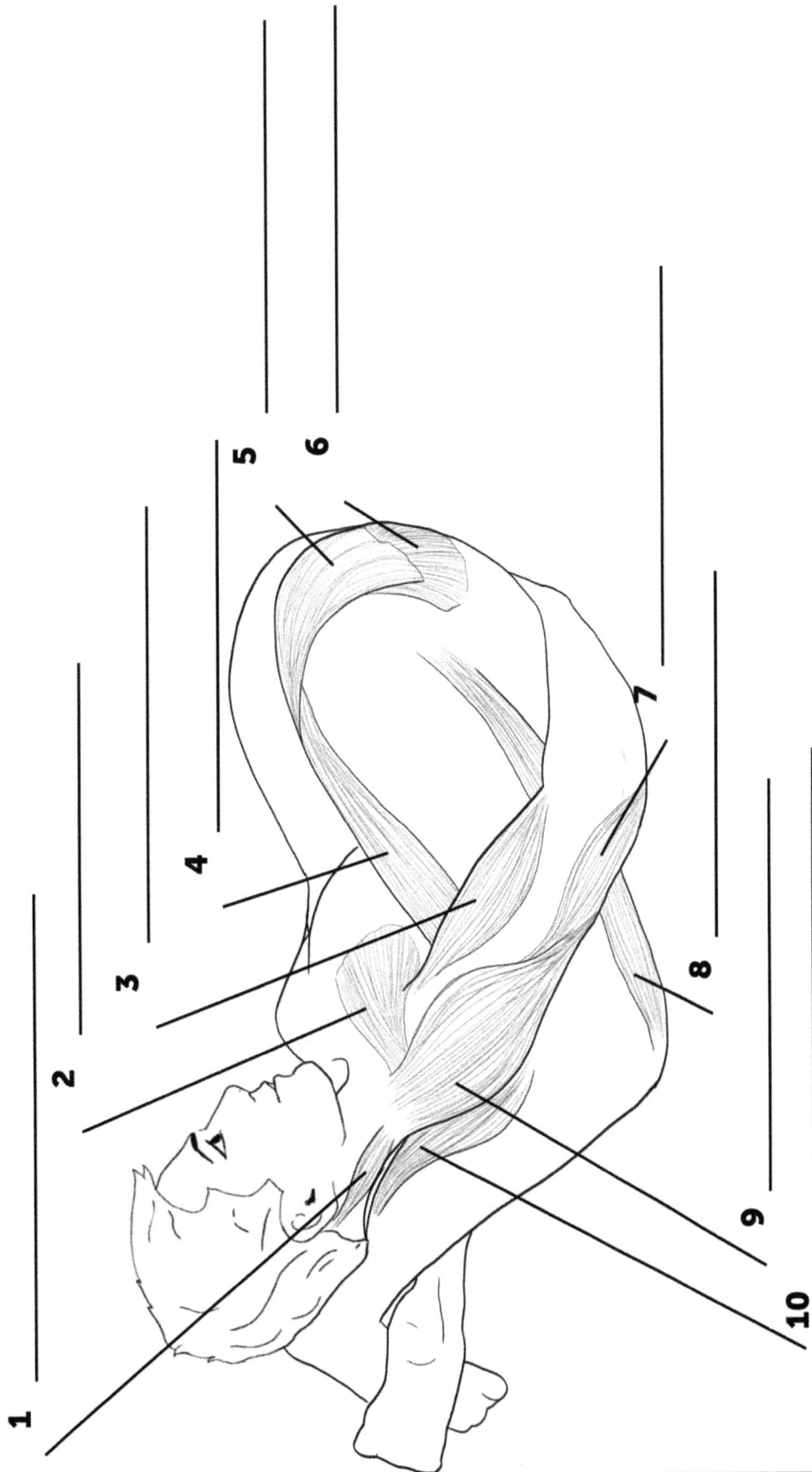

1

2

3

4

5

6

7

8

9

10

132. YOGANIDRASANA

1. ESTERNOCLEIDOMASTOIDEO
2. PECTORAL MAYOR
3. BÍCEPS BRAQUIAL
4. ISQUIOTIBIALES
5. MÚSCULO GLÚTEO MAYOR
6. GLÚTEO MEDIO
7. TRÍCEPS BRAQUIAL
8. CUADRÍCEPS
9. DELTOIDES
10. GASTROCNEMIO

133. POSTURA DE LA PALOMA

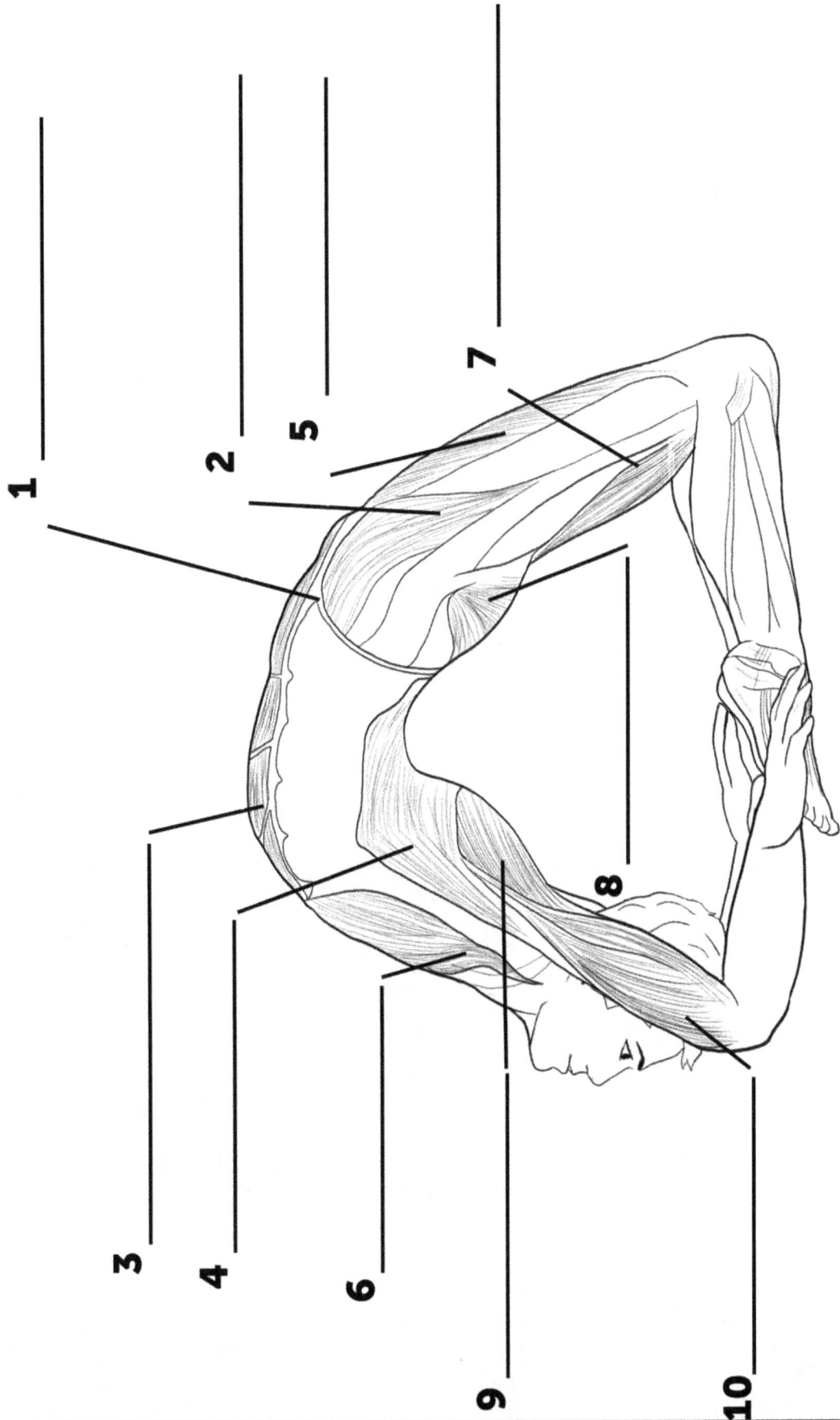

1

2

5

7

3

4

6

8

9

10

133. POSTURA DE LA PALOMA

1. ILIOPSOAS
2. MÚSCULO TENSOR DE LA FASCIA LATA
3. RECTO ABDOMINAL
4. LATISSIMUS DORSI
5. CUADRÍCEPS
6. PECTORAL MAYOR
7. ISQUIOTIBIALES
8. MÚSCULO GLÚTEO MAYOR
9. ERECTOR DE LA COLUMNA
10. TRÍCEPS BRAQUIAL

134. BADDHA KONA SIRSASANA

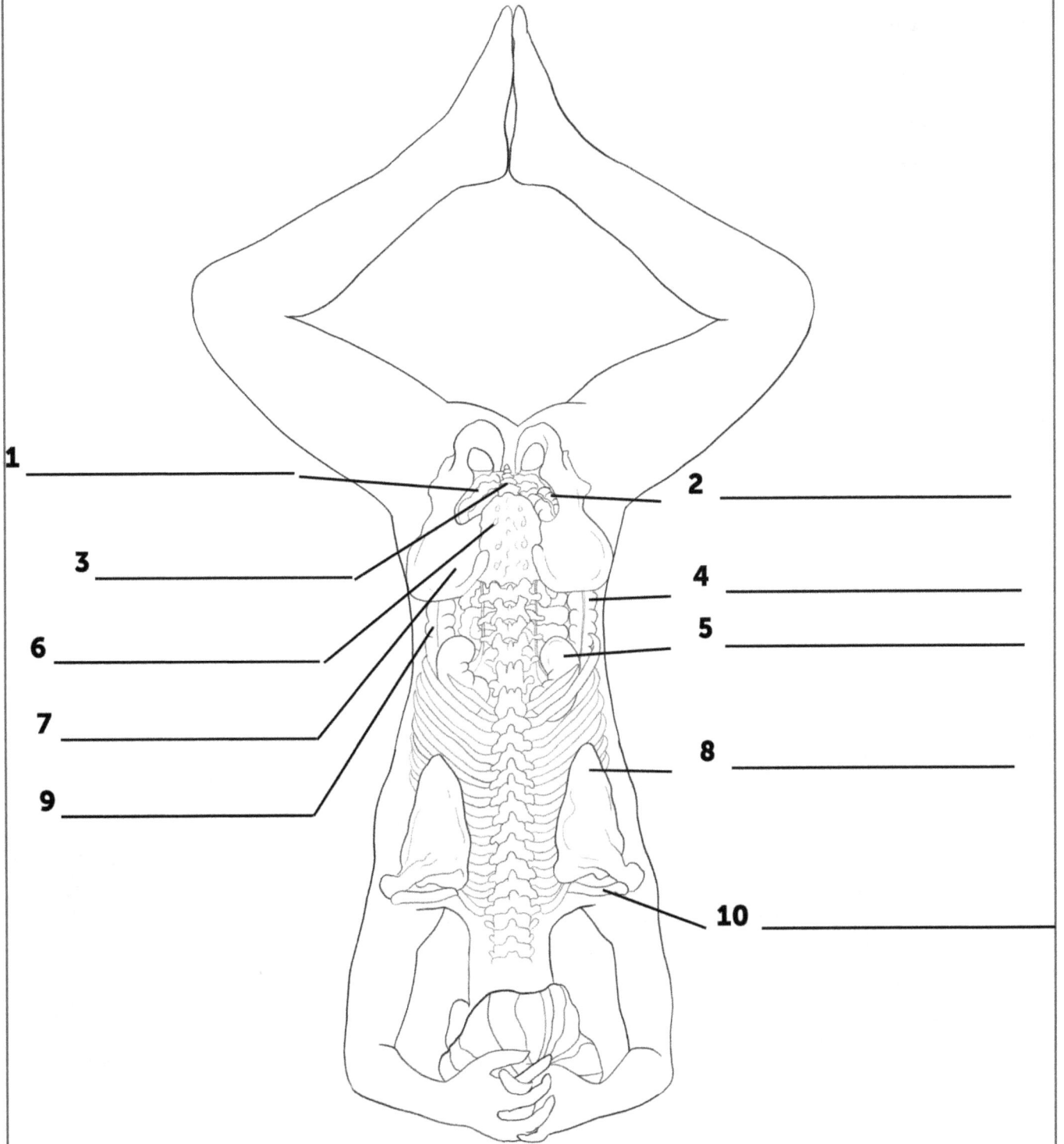

1 _____

2 _____

3 _____

4 _____

5 _____

6 _____

7 _____

8 _____

9 _____

10 _____

134. BADDHA KONA SIRSASANA

1. FOLICULOS DE INTESTINO DELGADO
2. SIGMOIDE
3. COXIS
4. COLON DESCENDENTE
5. RIÑÓN
6. SACRO
7. PELVIS
8. ESCÁPULA
9. COLON ASCENDENTE
10. CLAVÍCULA

135. VISVAMITRASANA II

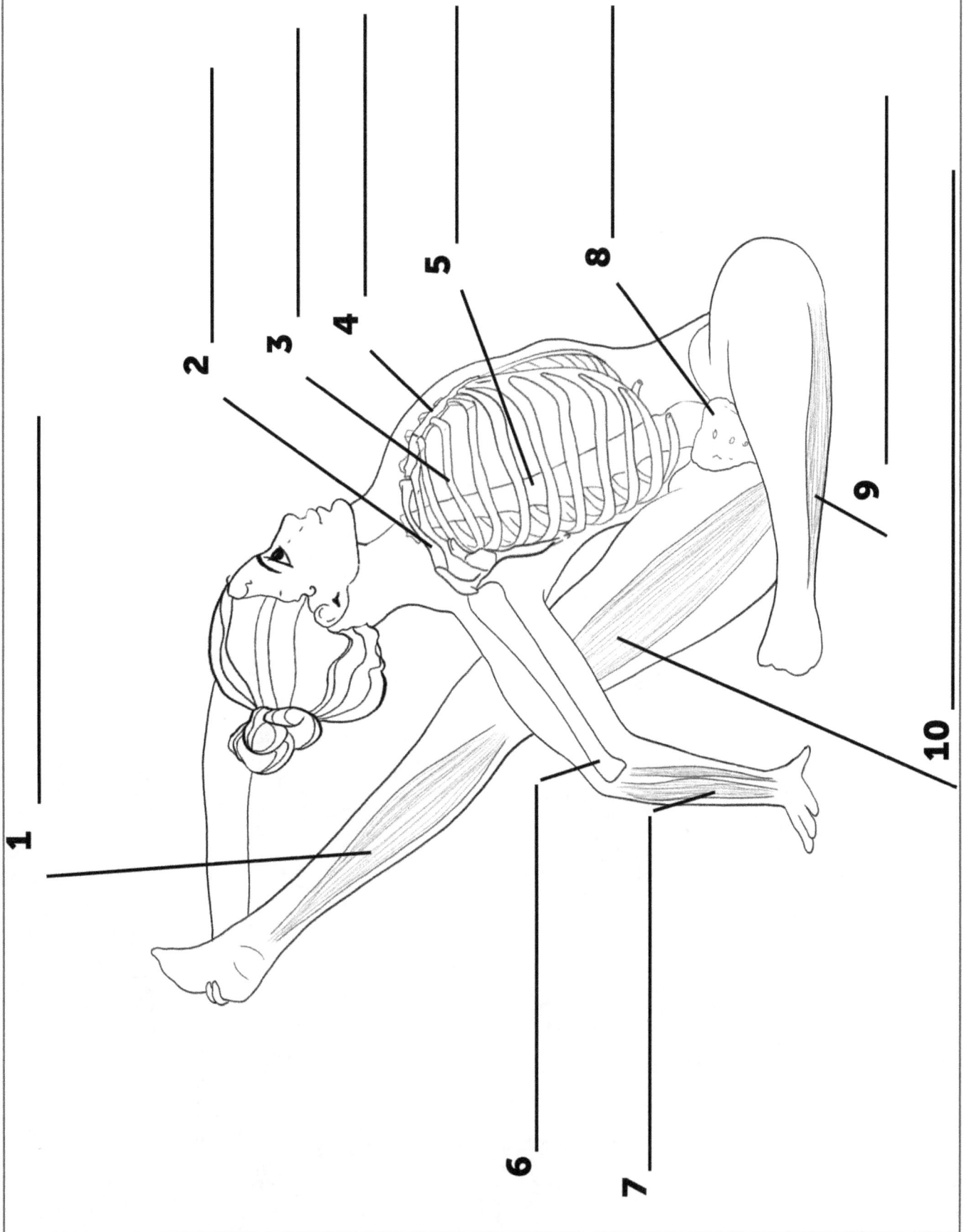

1

2

3

4

5

6

7

8

9

10

135. VISVAMITRASANA II

1. GASTROCNEMIO
2. CLAVÍCULA
3. COSTILLAS
4. ESTERNÓN
5. COLUMNA VERTEBRAL
6. HÚMERO
7. PRONADORES
8. SACRO
9. TIBIAL ANTERIOR
10. ISQUIOTIBIALES

136. POSTURA DE LOTO EN POSICIÓN DE HOMBRO

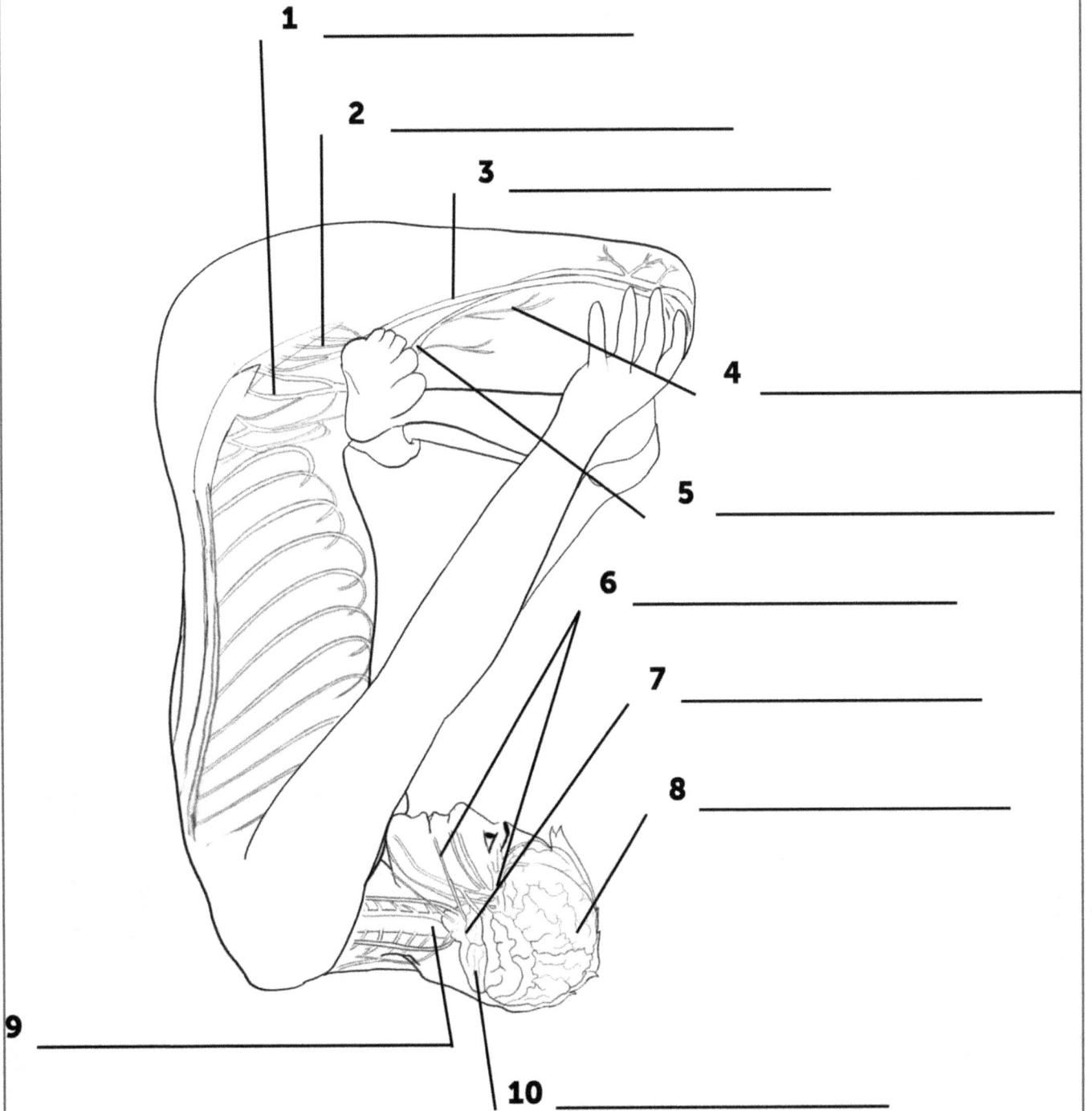

1 _____

2 _____

3 _____

4 _____

5 _____

6 _____

7 _____

8 _____

9 _____

10 _____

136. POSTURA DE LOTO EN POSICIÓN DE HOMBRO

1. PLEXO LUMBAR
2. PLEXO SACRO
3. CIÁTICO
4. RAMAS MUSCULARES DE FEMORAL
5. FEMORAL
6. NERVIOS CRANEALES
7. TRONCO ENCEFÁLICO
8. CEREBRO
9. MÉDULA ESPINAL
10. CEREBELO

137. POSTURA DE UNA RUEDA CON UNA PIERNA

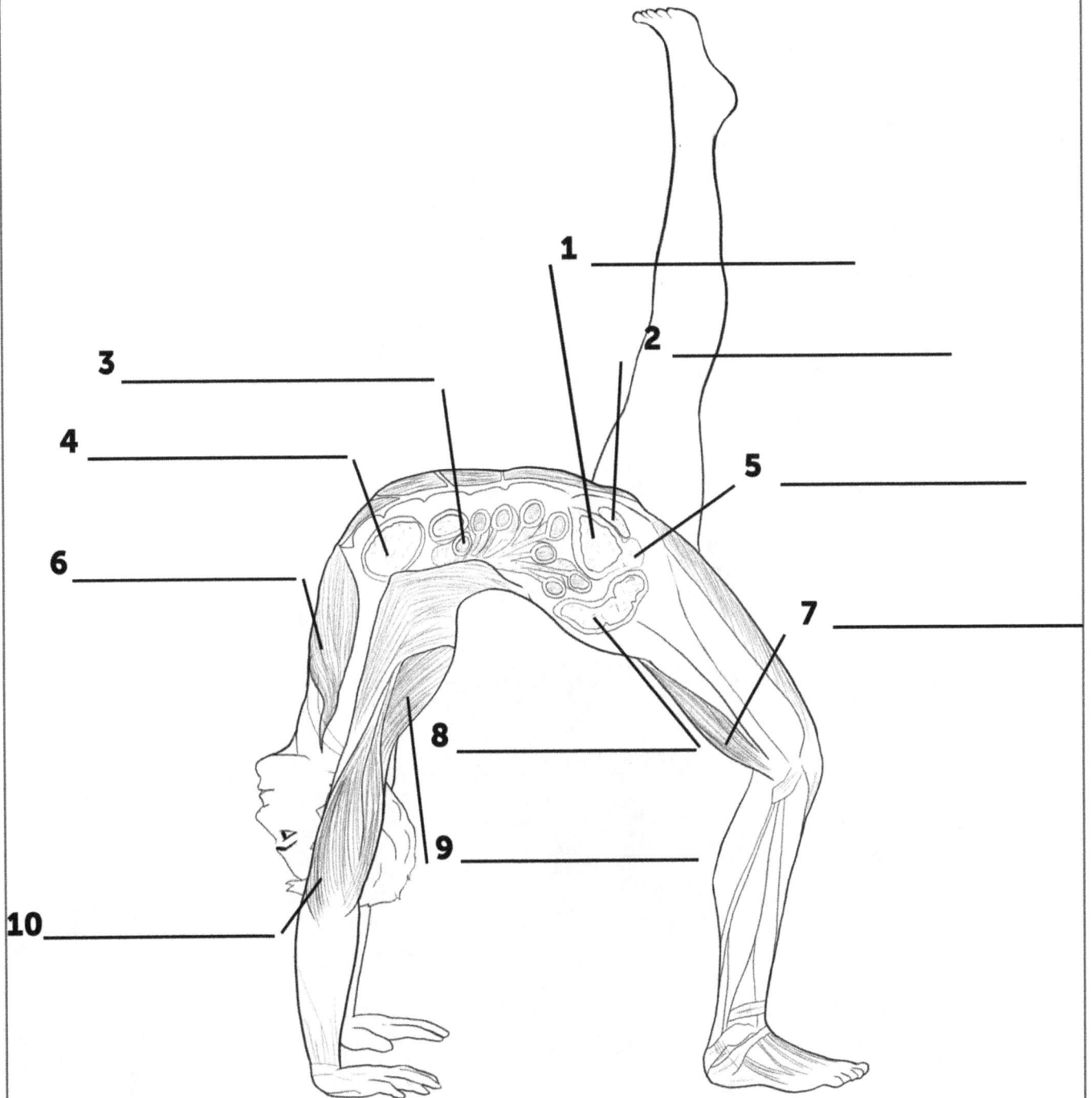

1 _____

2 _____

3 _____

4 _____

5 _____

6 _____

7 _____

8 _____

9 _____

10 _____

137. POSTURA DE UNA RUEDA CON UNA PIERNA

1. VEJIGA URINARIA

2. HUESO PÚBICO

3. FOLICULOS DE INTESTINO DELGADO

4. ESTÓMAGO

5. PRÓSTATA

6. PECTORAL MAYOR

7. ISQUIOTIBIALES

8. RECTO

9. ERECTOR DE LA COLUMNA

10. TRÍCEPS BRAQUIAL

138. EKA PADA SIRSASANA

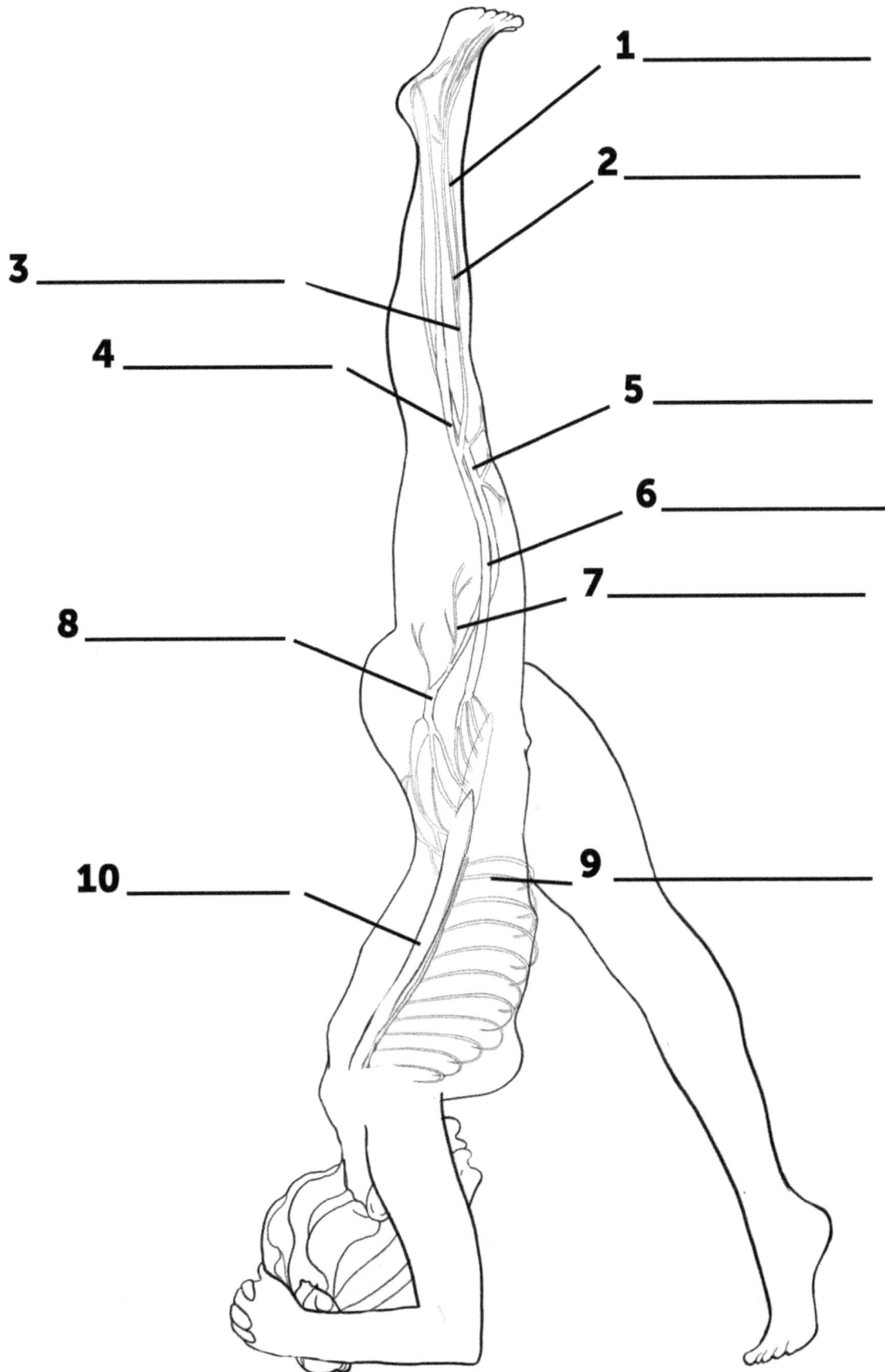

1 _____

2 _____

3 _____

4 _____

5 _____

6 _____

7 _____

8 _____

9 _____

10 _____

138. EKA PADA SIRSASANA

1. PERONEO SUPERFICIAL
2. PERONEO PROFUNDO
3. PERONEO COMÚN
4. TIBIAL
5. SAFENA
6. CIÁTICO
7. RAMAS MUSCULARES DE FEMORAL
8. FEMORAL
9. INTERCOSTALES
10. MÉDULA ESPINAL

139. SUPTA VISVAMITRASANA

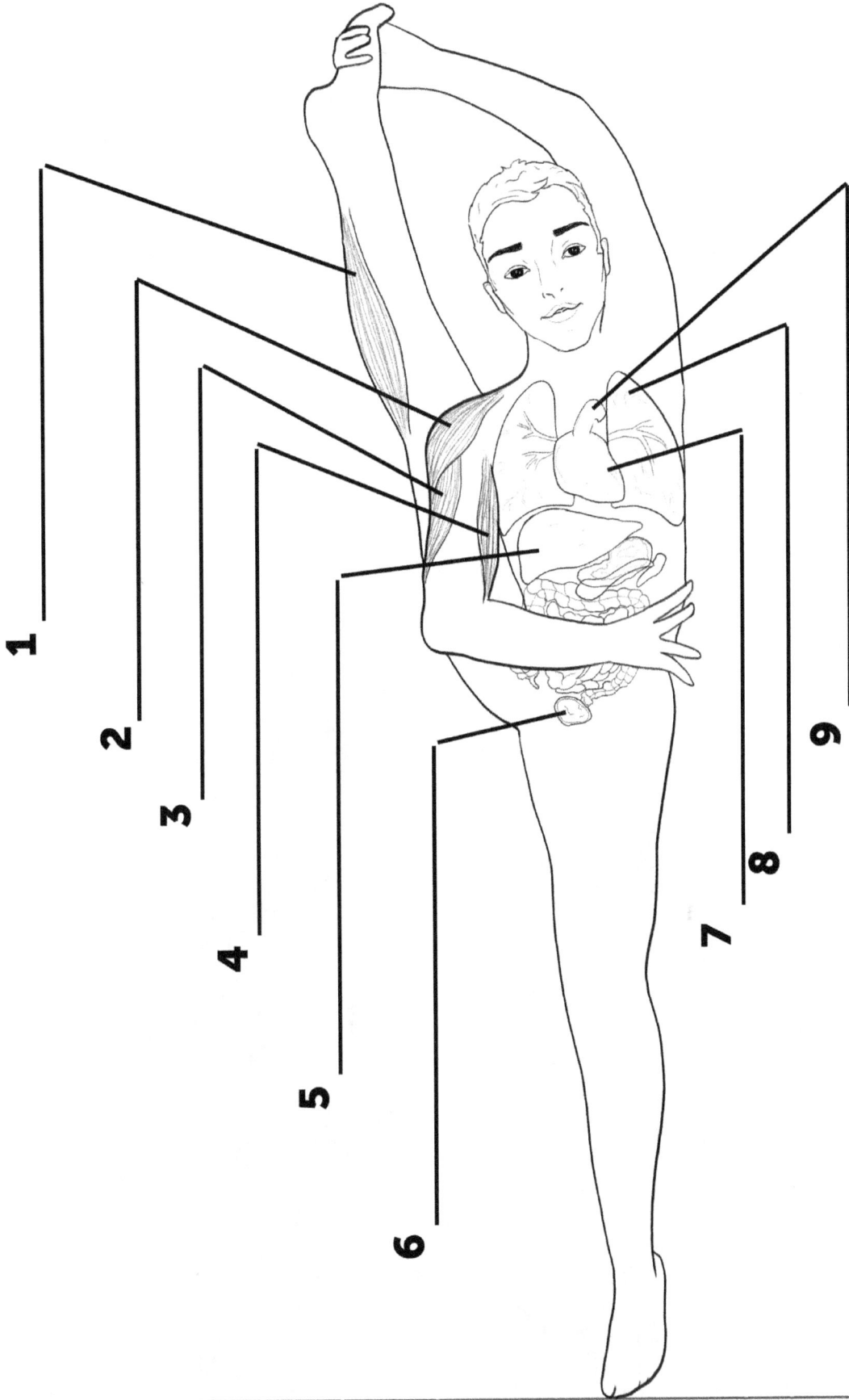

1
2
3
4
5
6
7
8
9

139. SUPTA VISVAMITRASANA

1. GASTROCNEMIO
2. DELTOIDES
3. TRÍCEPS BRAQUIAL
4. BÍCEPS BRAQUIAL
5. HÍGADO
6. VEJIGA URINARIA
7. CORAZÓN
8. PULMONES
9. AORTA

140. POSTURA DE FLEXIÓN HACIA ADELANTE MIRANDO HACIA ARRIBA

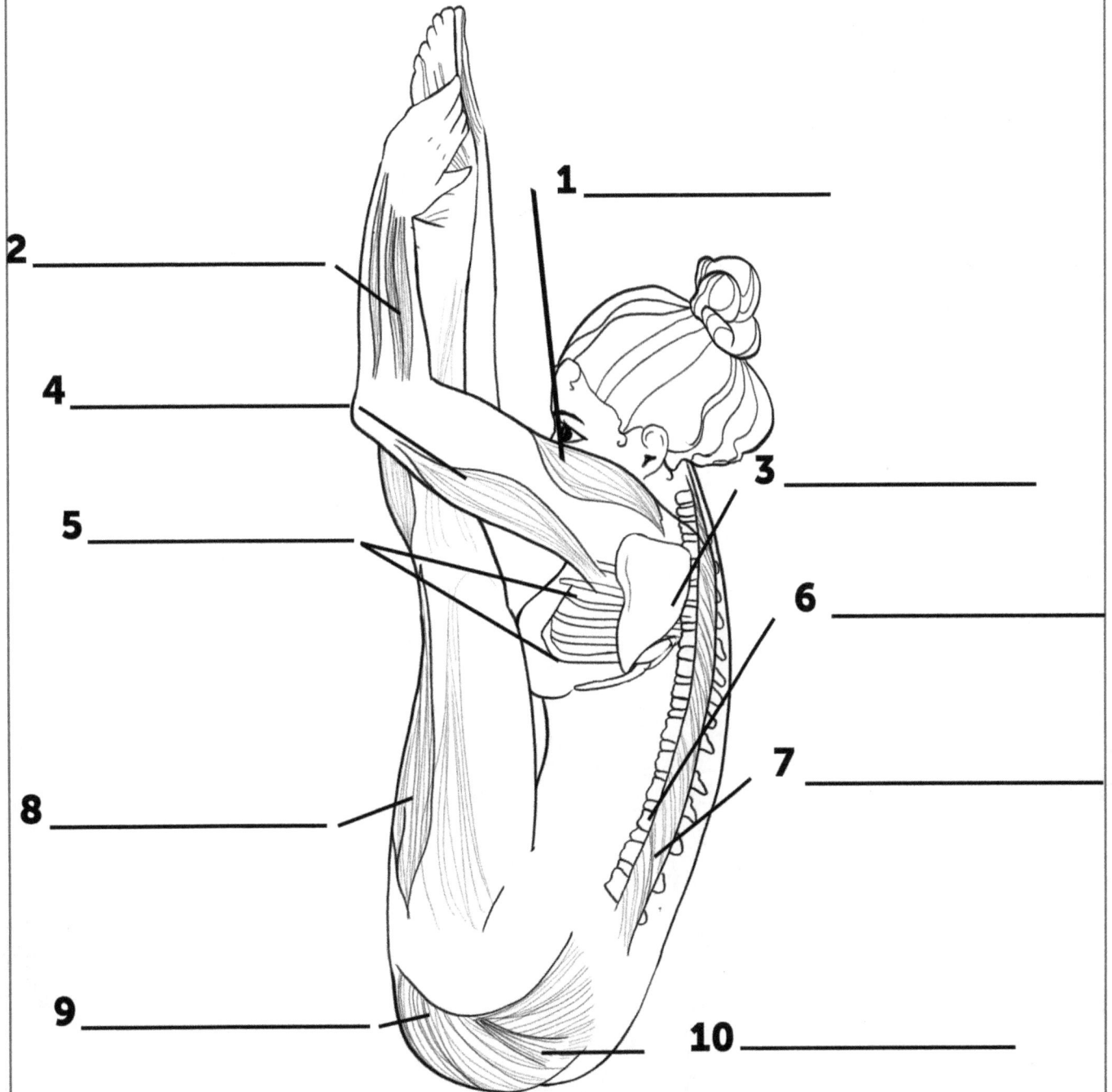

1 _____

2 _____

4 _____

3 _____

5 _____

6 _____

7 _____

8 _____

9 _____

10 _____

140. POSTURA DE FLEXIÓN HACIA ADELANTE MIRANDO HACIA ARRIBA

1. DELTOIDES

2. PRONADORES

3. ESCÁPULA

4. TRÍCEPS BRAQUIAL

5. COSTILLAS

6. COLUMNA VERTEBRAL

7. MÚSCULOS ESPINALES

8. ISQUIOTIBIALES

9. MÚSCULO GLÚTEO MAYOR

10. PIRIFORME

141. URDHVA UPAVISTHA KONASANA

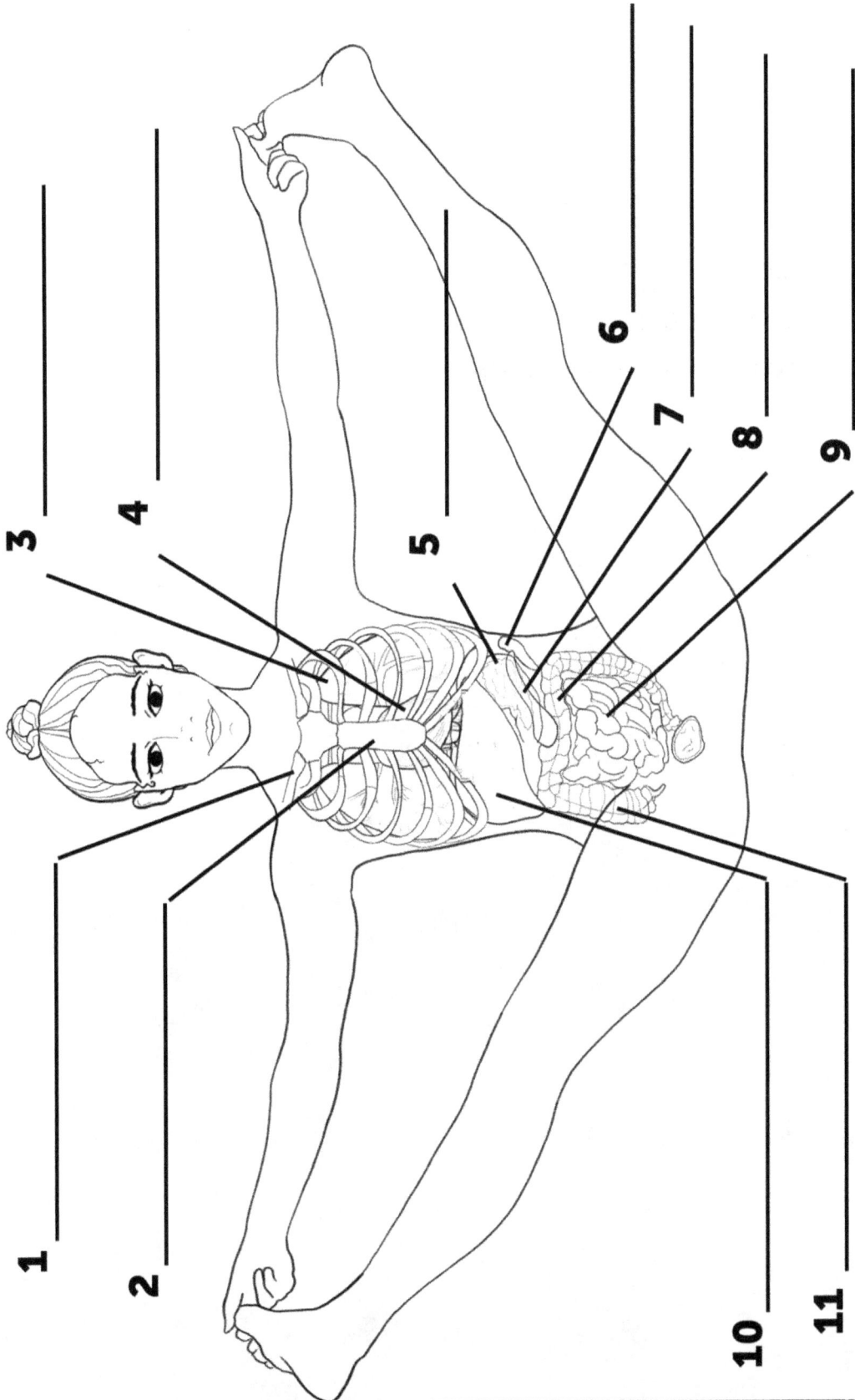

1

2

3

4

5

6

7

8

9

10

11

141. URDHVA UPAVISTHA KONASANA

1. CLAVÍCULA
2. ESTERNÓN
3. PULMONES
4. CORAZÓN
5. ESTÓMAGO
6. BAZO
7. PÁNCREAS
8. COLON TRANSVERSO
9. FOLICULOS DE INTESTINO DELGADO
10. HÍGADO
11. COLON ASCENDENTE

142 VISVAMITRASANA

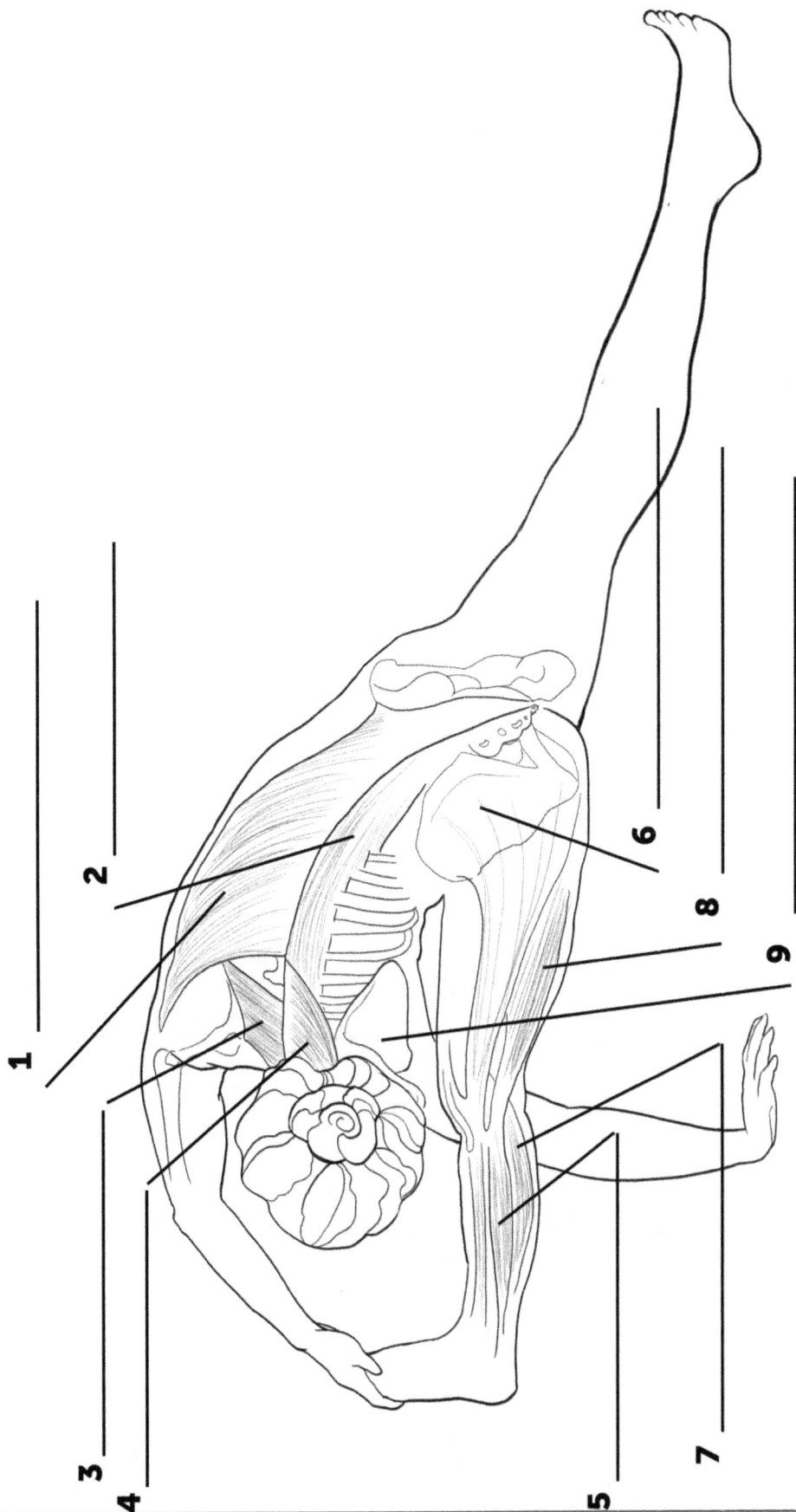

142 VISVAMITRASANA

1. LATISSIMUS DORSI

2. ERECTOR DE LA COLUMNA

3. ROMBOIDES

4. TRAPECIO

5. SÓLEO

6. PELVIS

7. GASTROCNEMIO

8. ISQUIOTIBIALES

9. ESCÁPULA

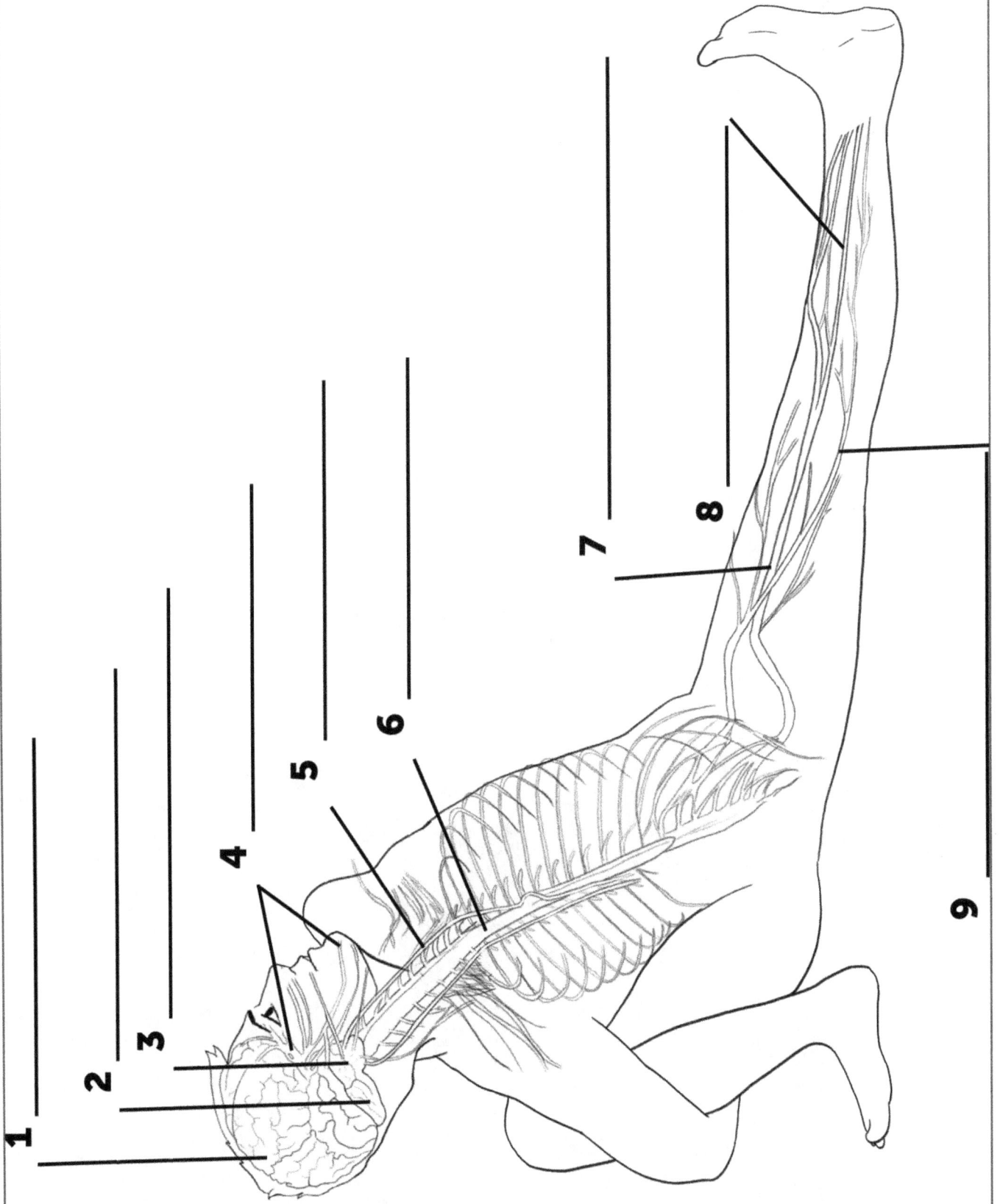

1

2

3

4

5

6

7

8

9

143 UTTHITA BADDHA PARSVA UPAVESASANA

1. CEREBRO
2. CEREBELO
3. TRONCO ENCEFÁLICO
4. NERVIOS CRANEALES
5. NERVIO VAGO
6. MÉDULA ESPINAL
7. CIÁTICO
8. TIBIAL
9. SAFENA

144. BHAKTI VIRABHADRASANA

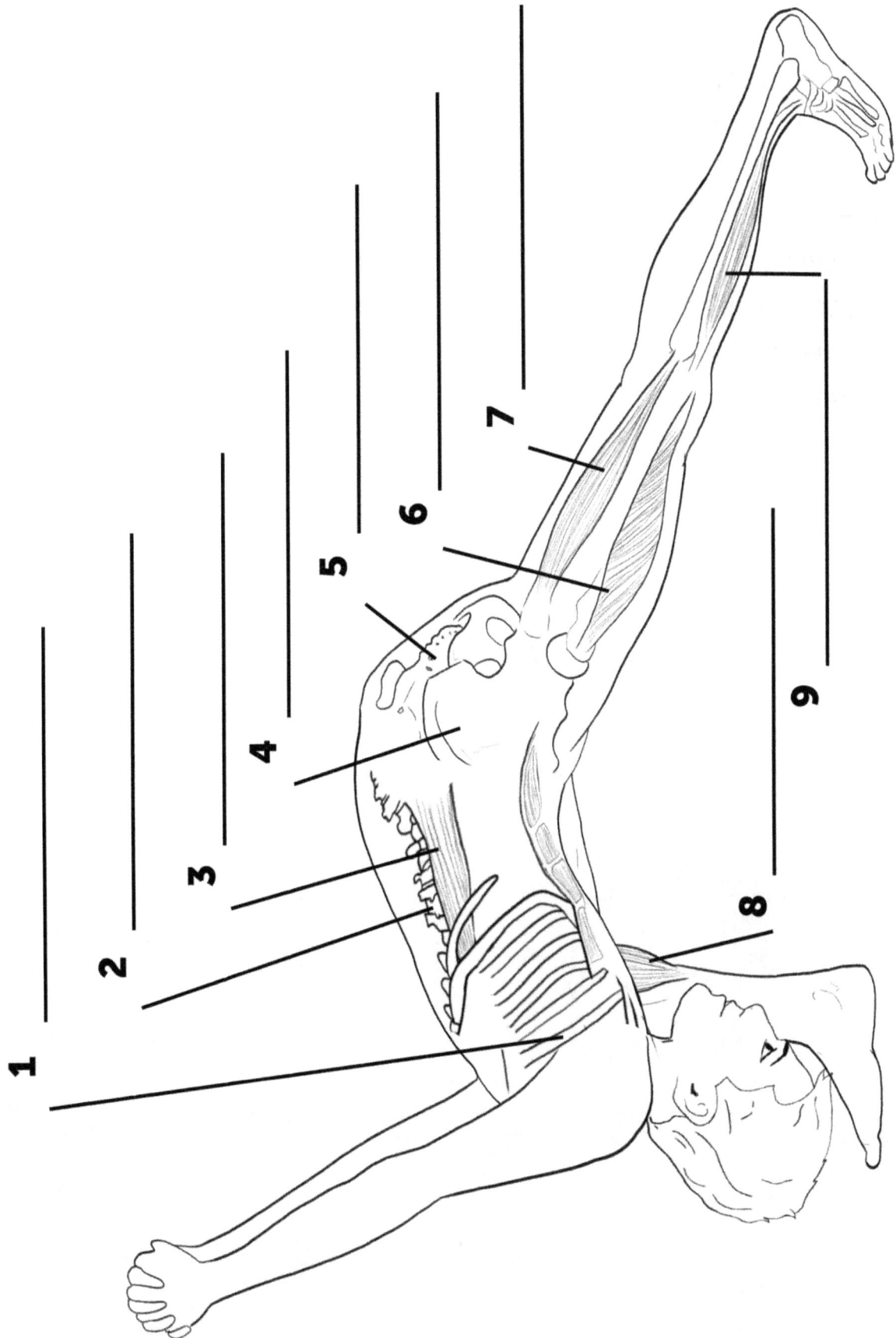

144. BHAKTI VIRABHADRASANA

1. COSTILLAS
2. COLUMNA VERTEBRAL
3. ERECTOR DE LA COLUMNA
4. PELVIS
5. SACRO
6. CUADRÍCEPS
7. ISQUIOTIBIALES
8. GASTROCNEMIO
9. TIBIAL ANTERIOR

145. BADDHA UTTHAN PRISTHASANA

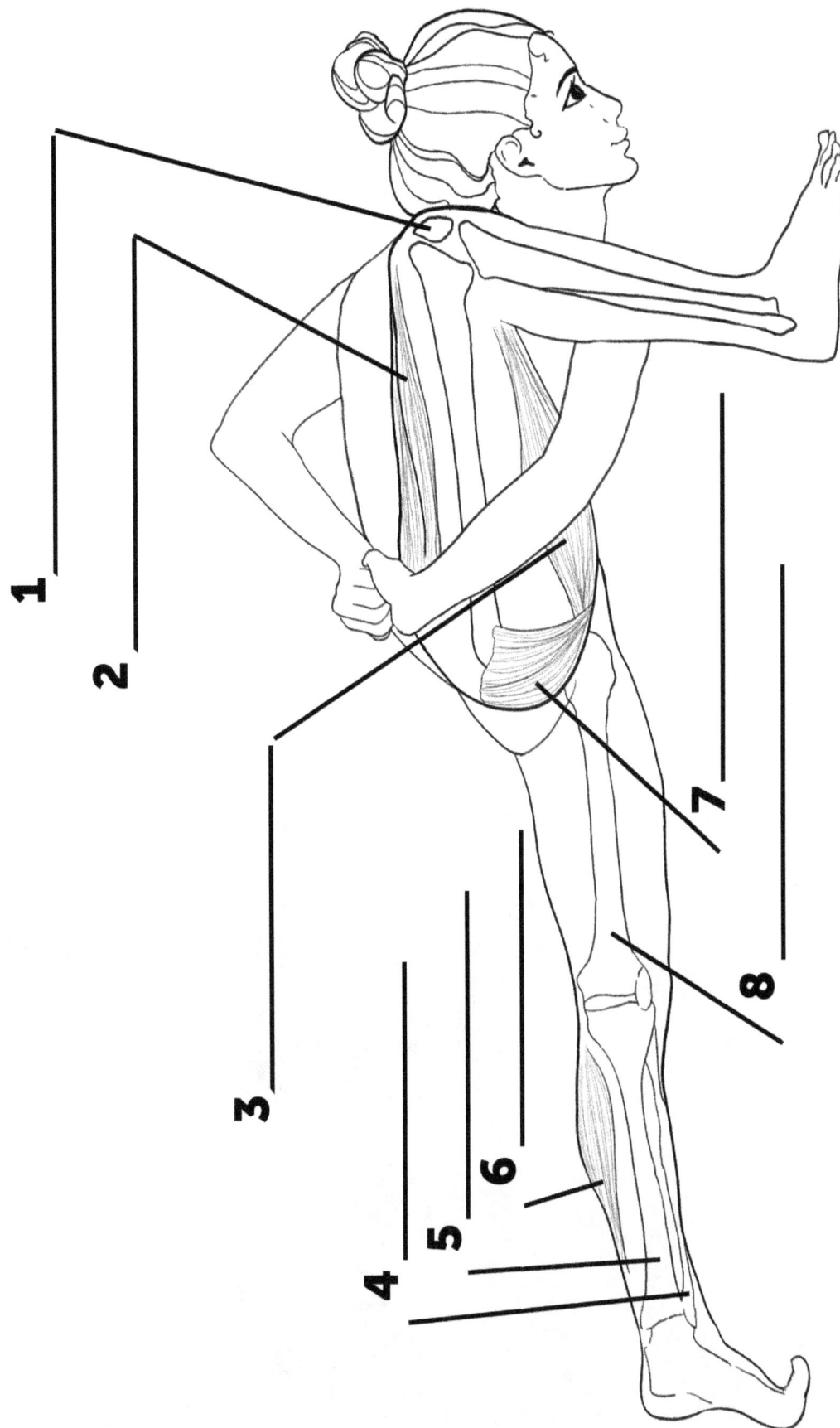

1

2

3

4

5

6

7

8

145. BADDHA UTTHAN PRISTHASANA

1. RÓTULA
2. CUADRÍCEPS
3. ISQUIOTIBIALES
4. PERONÉ
5. TIBIA
6. GASTROCNEMIO
7. MÚSCULO GLÚTEO MAYOR
8. FÉMUR

146. URDHVA PRASARITA EKA PADASANA

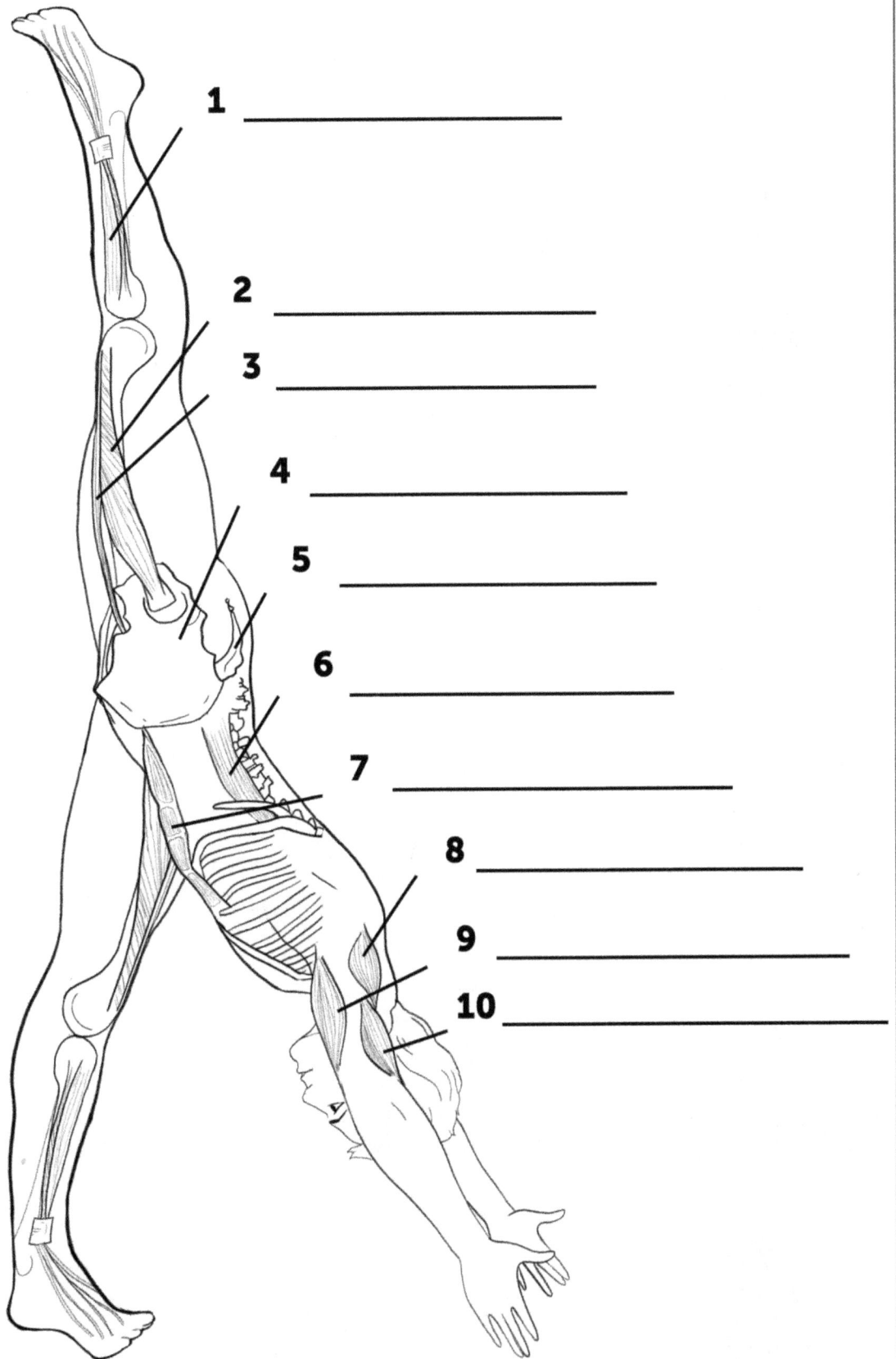

1 _____

2 _____

3 _____

4 _____

5 _____

6 _____

7 _____

8 _____

9 _____

10 _____

146. URDHVA PRASARITA EKA PADASANA

1.   TIBIAL ANTERIOR
2.   RECTO FEMORAL
3.   SARTORIO
4.   PELVIS
5.   SACRO
6.   ERECTOR DE LA COLUMNA
7.   RECTO ABDOMINAL
8.   DELTOIDES
9.   BÍCEPS BRAQUIAL
10. TRÍCEPS BRAQUIAL

147. VIRABHADRASANA III

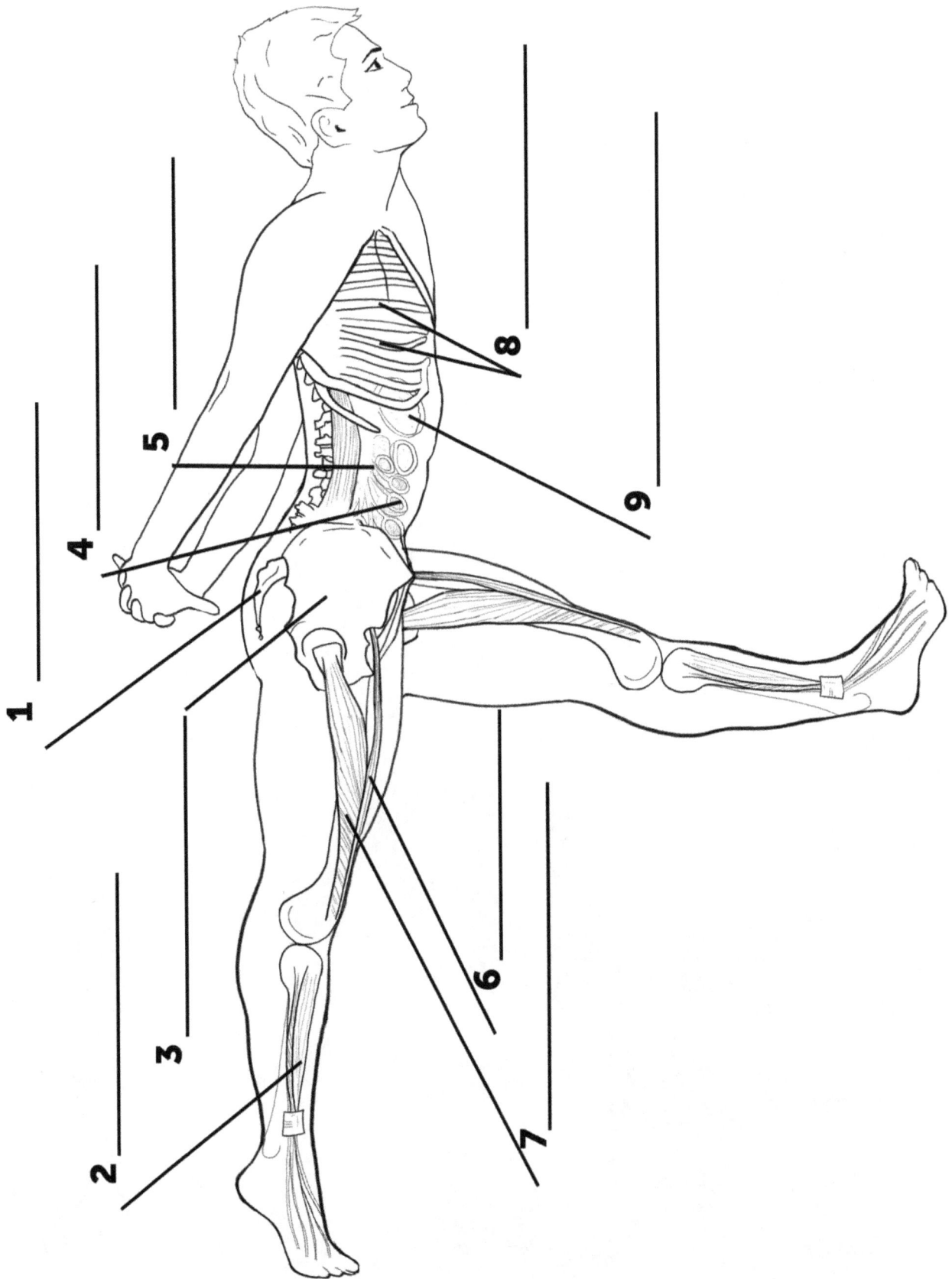

147. VIRABHADRASANA III

1. SACRO
2. TIBIAL ANTERIOR
3. PELVIS
4. FOLICULOS DE INTESTINO DELGADO
5. MESENTERIO DEL INTESTINO DELGADO
6. SARTORIO
7. RECTO FEMORAL
8. COSTILLAS
9. ESTÓMAGO

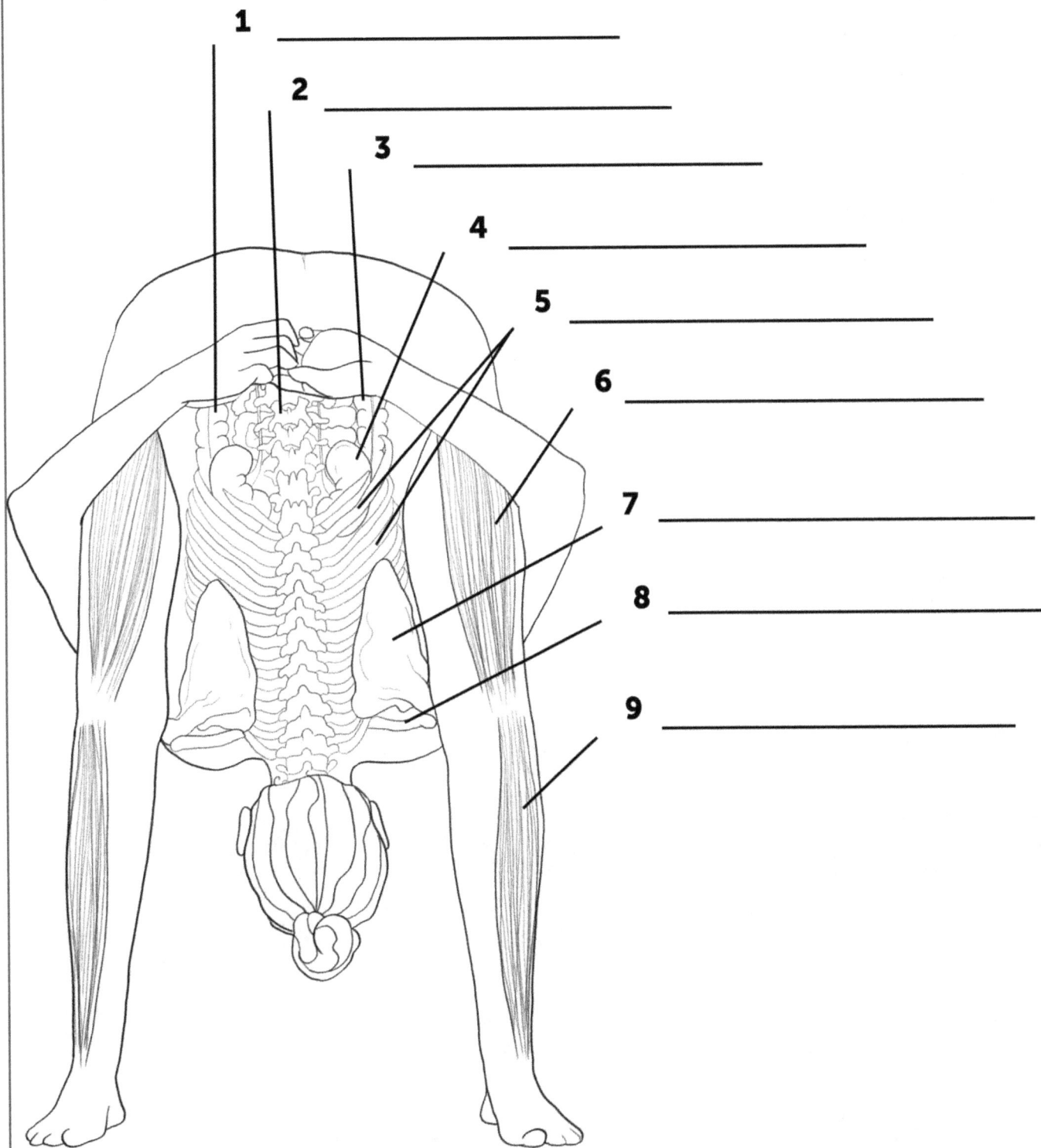

1 _____

2 _____

3 _____

4 _____

5 _____

6 _____

7 _____

8 _____

9 _____

148 PLIEGUE HACIA ADELANTE OBLIGADO

1. COLON ASCENDENTE
2. COLUMNA VERTEBRAL
3. COLON DESCENDENTE
4. RIÑÓN
5. COSTILLAS
6. CUADRÍCEPS
7. ESCÁPULA
8. CLAVÍCULA
9. TIBIAL ANTERIOR

149 UTTĀNĀSANA

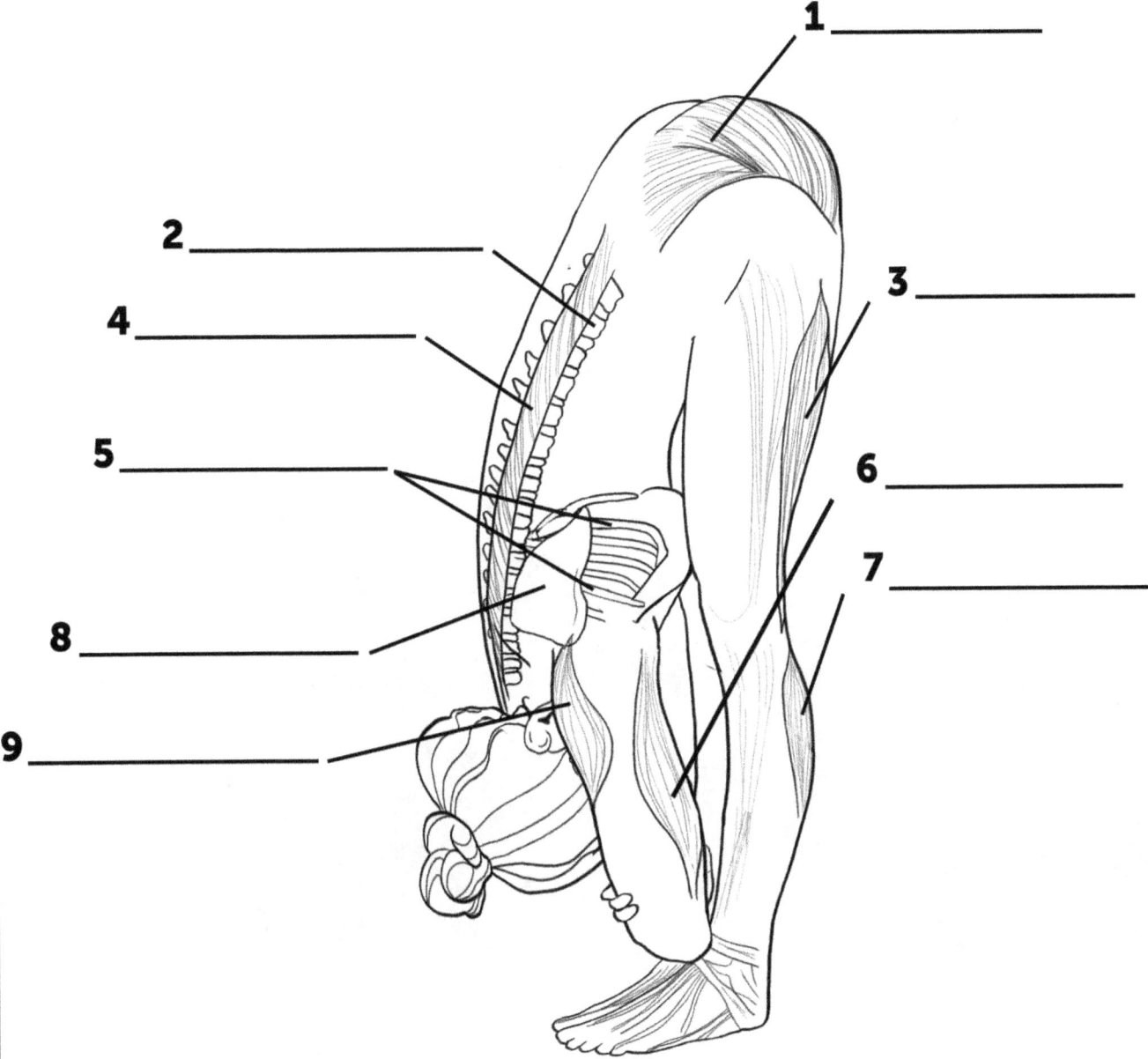

1 _____

2 _____

3 _____

4 _____

5 _____

6 _____

7 _____

8 _____

9 _____

149   UTTāNāSANA

1. PIRIFORME

2. COLUMNA VERTEBRAL

3. ISQUIOTIBIALES

4. MÚSCULOS ESPINALES

5. COSTILLAS

6. TRÍCEPS BRAQUIAL

7. GASTROCNEMIO

8. ESCÁPULA

9. DELTOIDES

150. POSTURA DE MEDIA PALOMA DESCANSADA

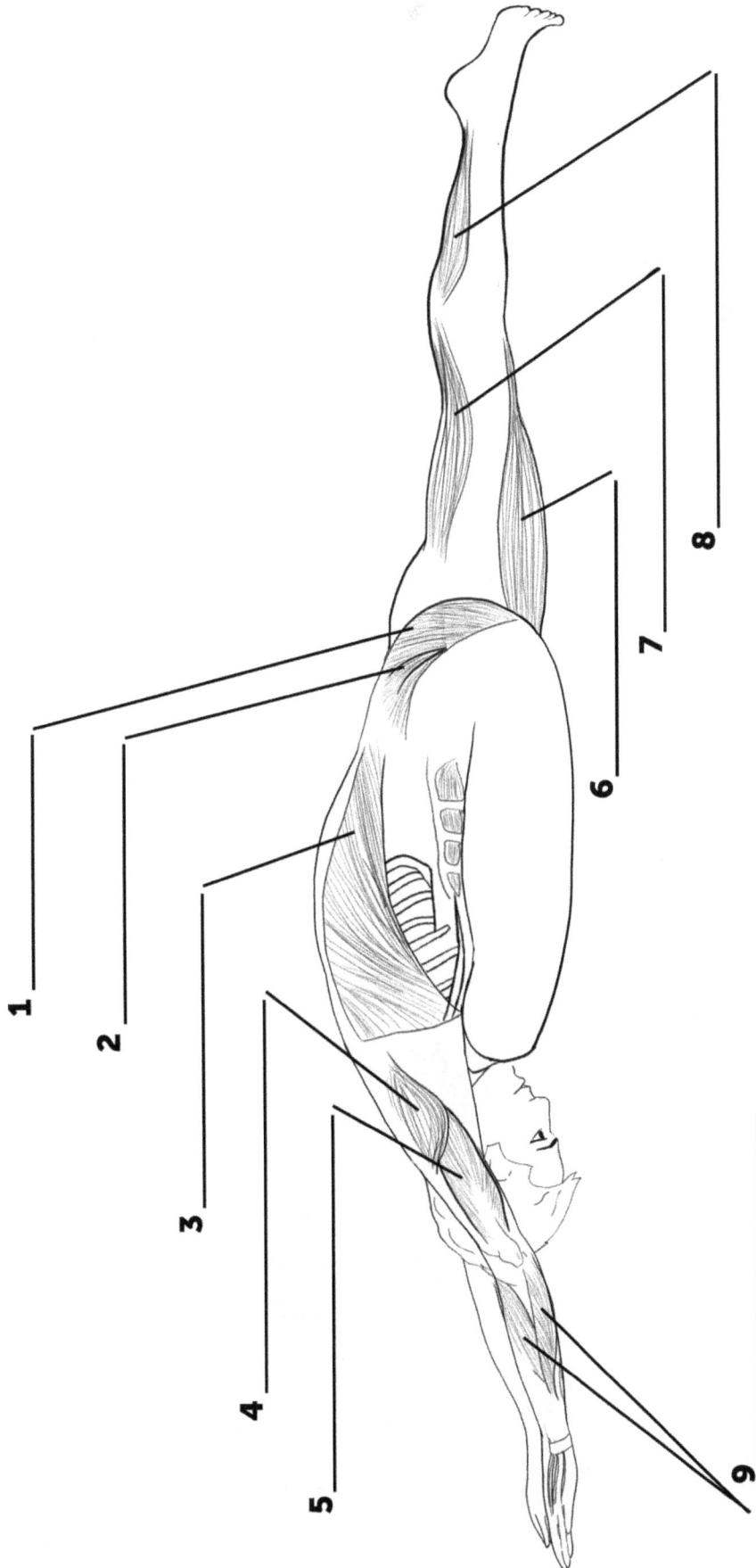

1

2

3

4

5

6

7

8

9

150. POSTURA DE MEDIA PALOMA DESCANSADA

1. MÚSCULO GLÚTEO MAYOR
2. PIRIFORME
3. LATISSIMUS DORSI
4. DELTOIDES
5. TRÍCEPS BRAQUIAL
6. CUADRÍCEPS
7. ISQUIOTIBIALES
8. GASTROCNEMIO
9. PRONADORES

151. EKA PADA ARDHA PURVOTTANASANA

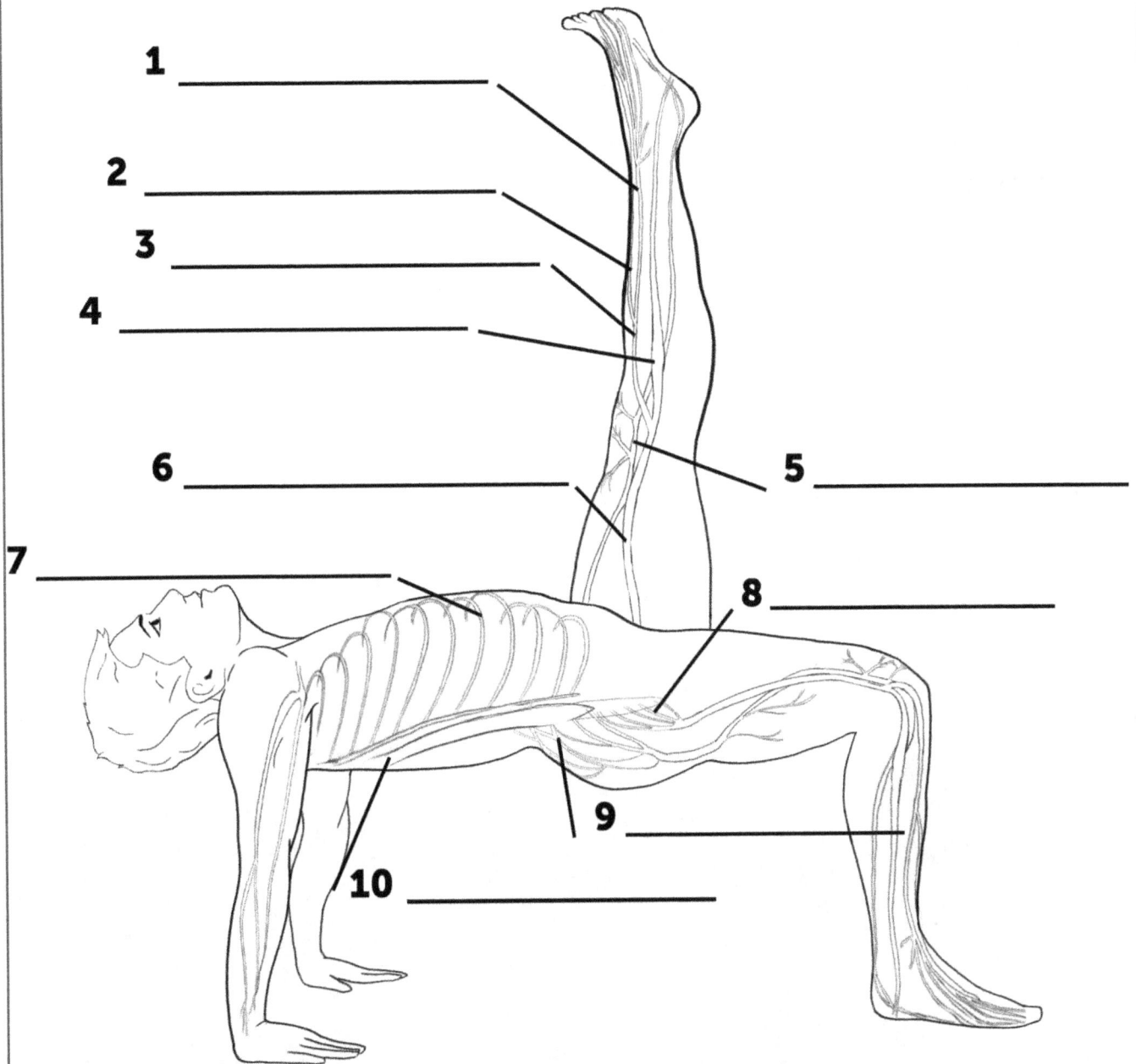

1 _____

2 _____

3 _____

4 _____

6 _____

7 _____

5 _____

8 _____

9 _____

10 _____

151. EKA PADA ARDHA PURVOTTANASANA

1. PERONEO PROFUNDO
2. PERONEO SUPERFICIAL
3. PERONEO COMÚN
4. TIBIAL
5. SAFENA
6. CIÁTICO
7. INTERCOSTALES
8. PLEXO SACRO
9. PLEXO LUMBAR
10. MÉDULA ESPINAL

152. EKA PADA BAKASANA II

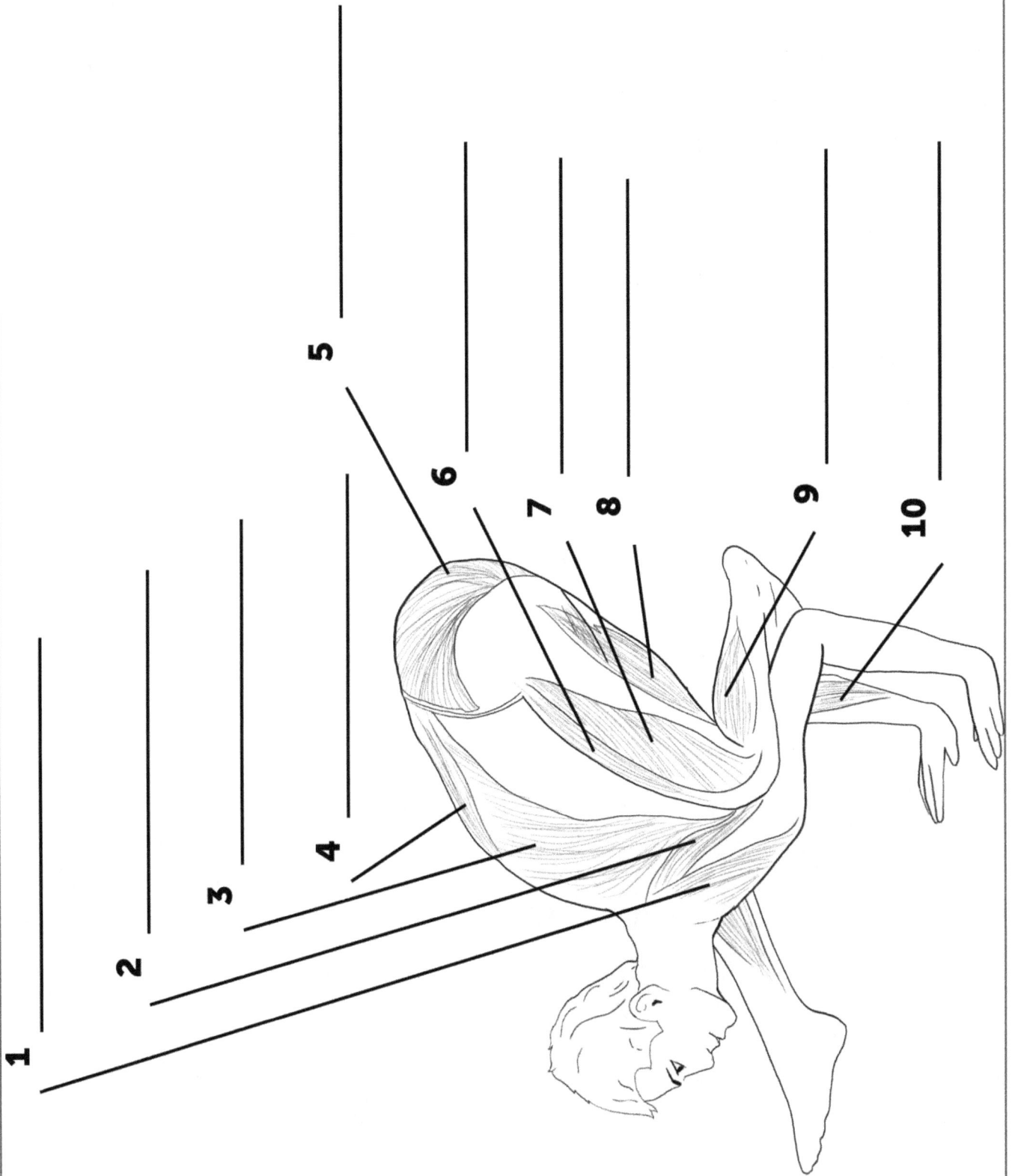

1

2

3

4

5

6

7

8

9

10

152. EKA PADA BAKASANA II

1. DELTOIDES
2. TRÍCEPS BRAQUIAL
3. LATISSIMUS DORSI
4. ERECTOR DE LA COLUMNA
5. MÚSCULO GLÚTEO MAYOR
6. RECTO FEMORAL
7. MÚSCULO VASTO LATERAL
8. ISQUIOTIBIALES
9. GASTROCNEMIO
10. PRONADORES

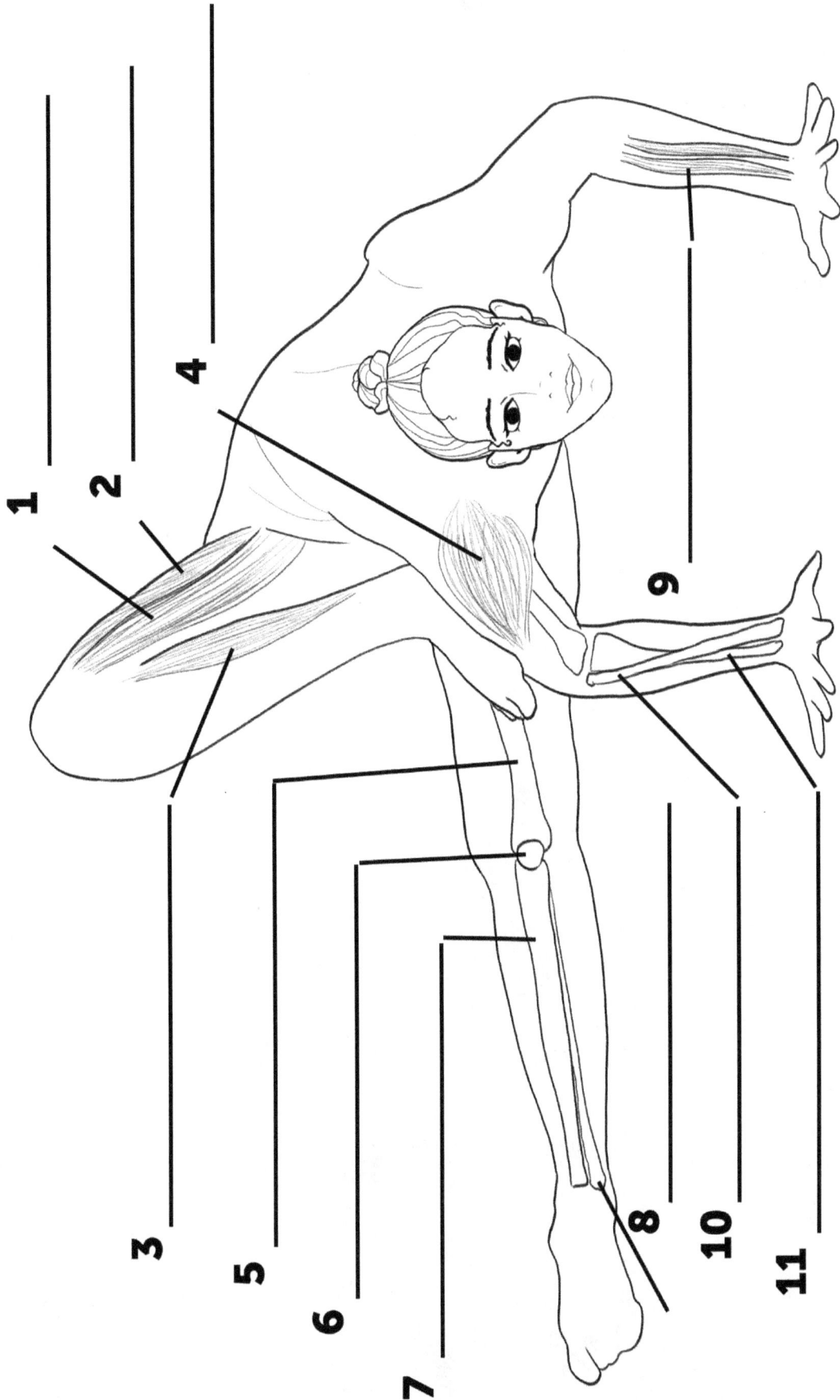

153 LIBÉLULA

1

2

3

4

5

6

7

8

9

10

11

153 LIBÉLULA

1. MÚSCULO VASTO LATERAL
2. RECTO FEMORAL
3. GASTROCNEMIO
4. DELTOIDES
5. FÉMUR
6. RÓTULA
7. TIBIA
8. PERONÉ
9. PRONADORES
10. RADIO
11. CÚBITO

154. POSTURA DEL ÁRBOL CON UNA MANO

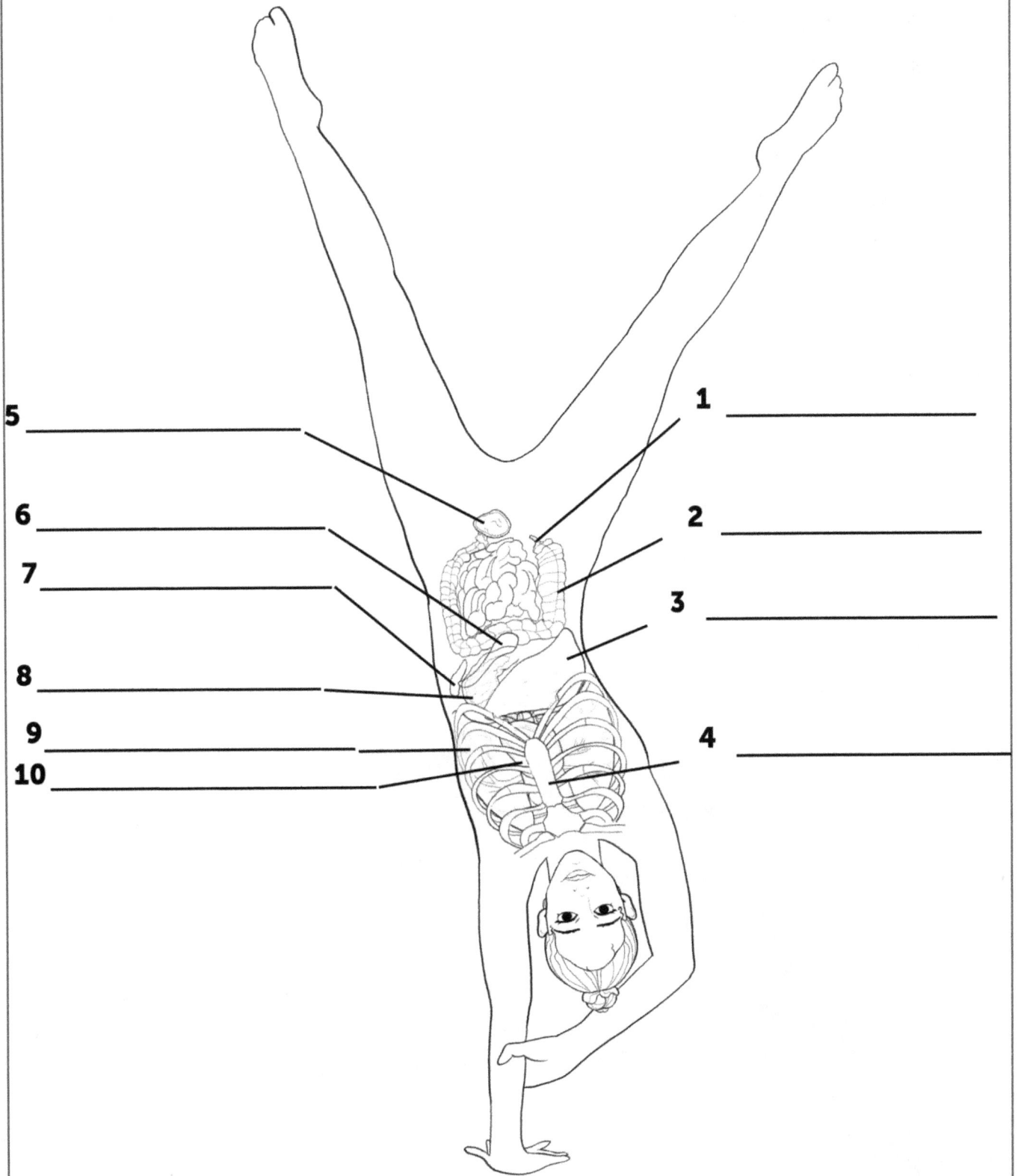

5 _____

6 _____

7 _____

8 _____

9 _____

10 _____

1 _____

2 _____

3 _____

4 _____

154. POSTURA DEL ÁRBOL CON UNA MANO

1. APÉNDICE

2. COLON ASCENDENTE

3. HÍGADO

4. ESTERNÓN

5. VEJIGA URINARIA

6. PÁNCREAS

7. BAZO

8. ESTÓMAGO

9. PULMONES

10. CORAZÓN

155 COBRA REAL

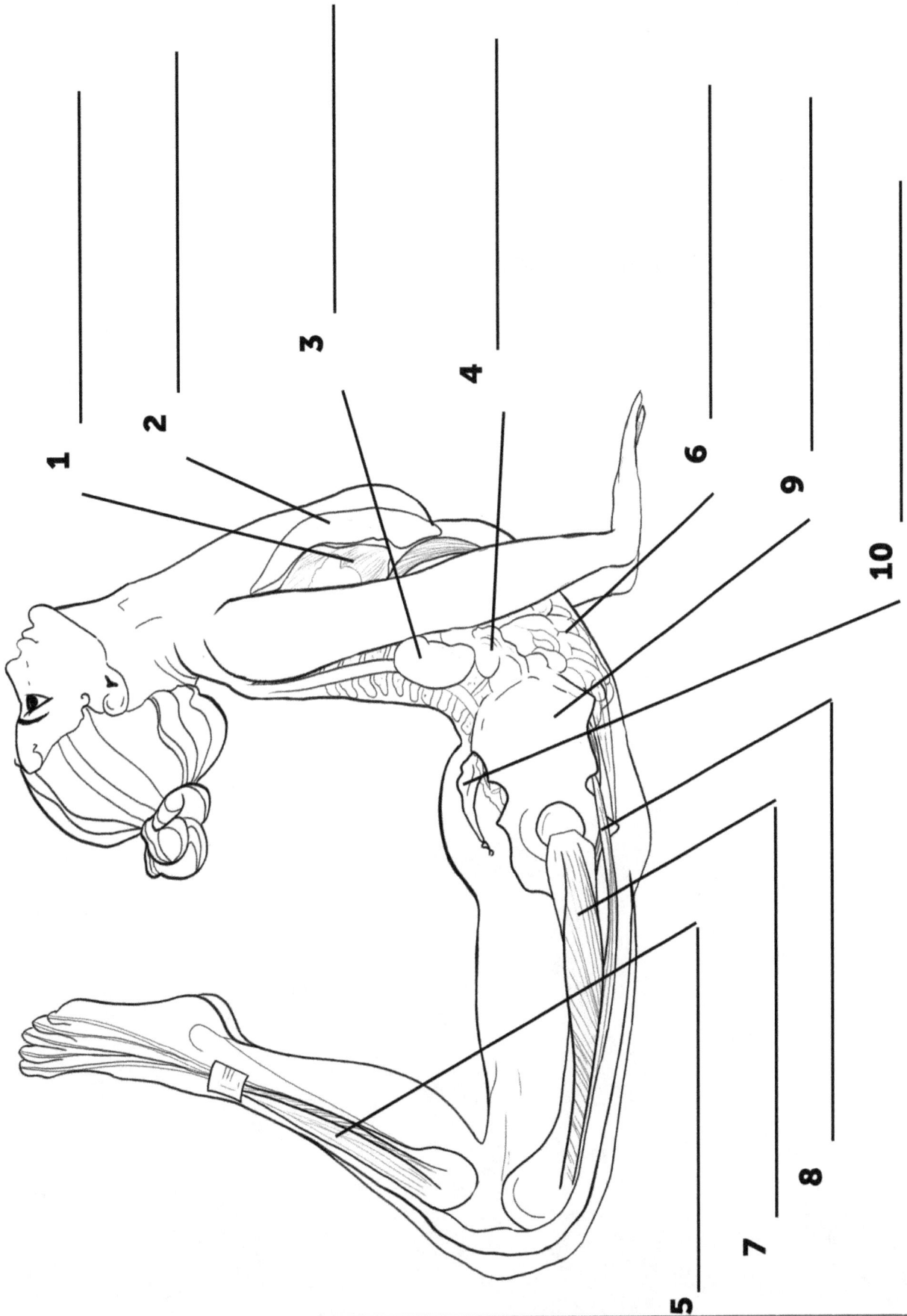

155 COBRA REAL

1. CORAZÓN

2. PULMONES

3. RIÑÓN

4. COLON ASCENDENTE

5. TIBIAL ANTERIOR

6. FOLICULOS DE INTESTINO DELGADO

7. RECTO FEMORAL

8. SARTORIO

9. PELVIS

10. SACRO

156. UTKATASANA

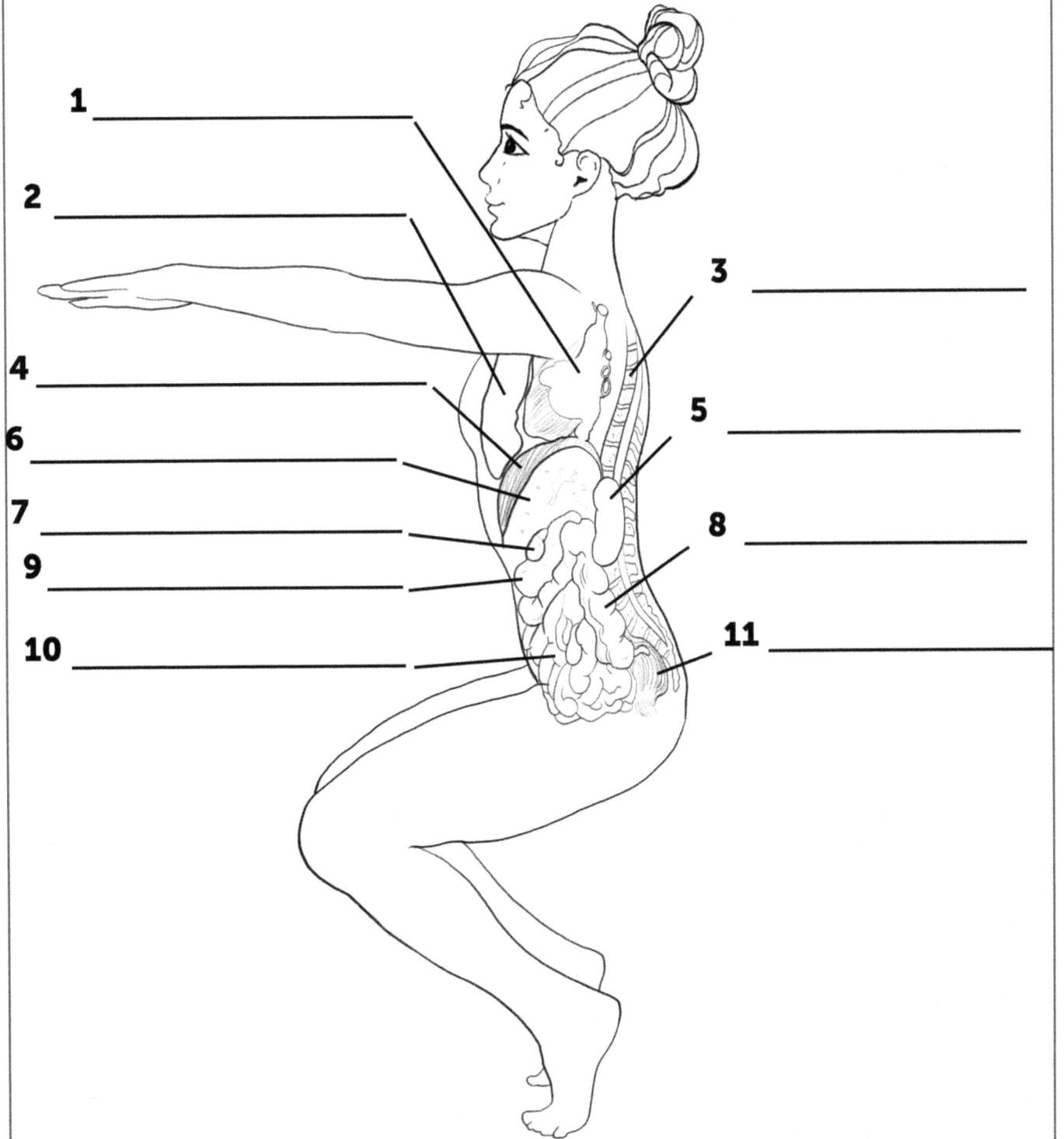

1 _____

2 _____

3 _____

4 _____

5 _____

6 _____

7 _____

8 _____

9 _____

10 _____

11 _____

156. UTKATASANA

1. CORAZÓN

2. PULMONES

3. COLUMNA VERTEBRAL

4. DIAFRAGMA

5. RIÑÓN

6. HÍGADO

7. VESÍCULA BILIAR

8. COLON DESCENDENTE

9. ESTÓMAGO

10. FOLICULOS DE INTESTINO DELGADO

11. RECTO

157 DANDAYAMANA JANUSHIRASANA

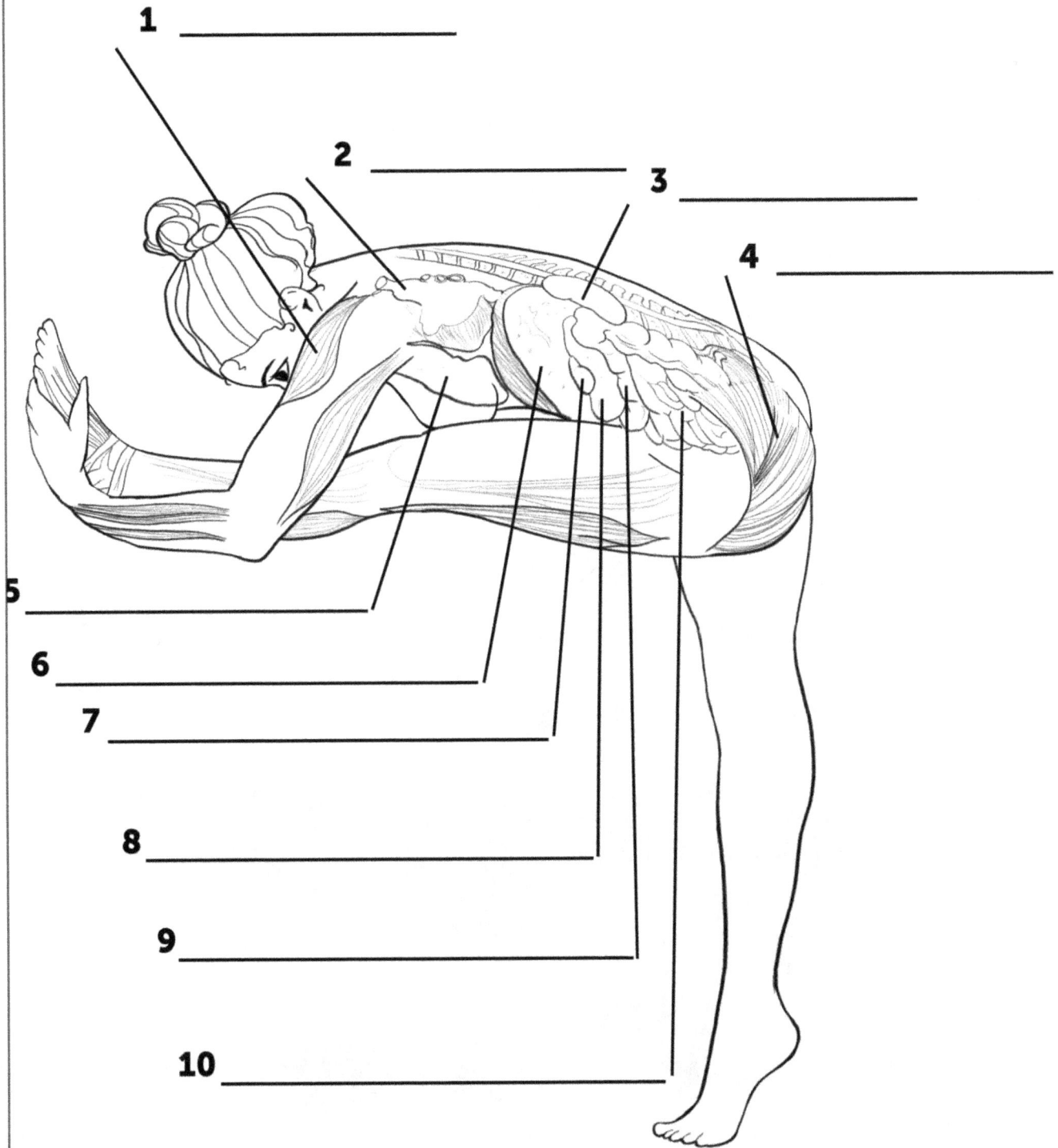

1 _____

2 _____

3 _____

4 _____

5 _____

6 _____

7 _____

8 _____

9 _____

10 _____

157 DANDAYAMANA JANUSHIRASANA

1. DELTOIDES

2. CORAZÓN

3. RIÑÓN

4. PIRIFORME

5. PULMONES

6. HÍGADO

7. VESÍCULA BILIAR

8. ESTÓMAGO

9. COLON TRANSVERSO

10. FOLICULOS DE INTESTINO DELGADO

158. NIRALAMBA SARVANGASANA

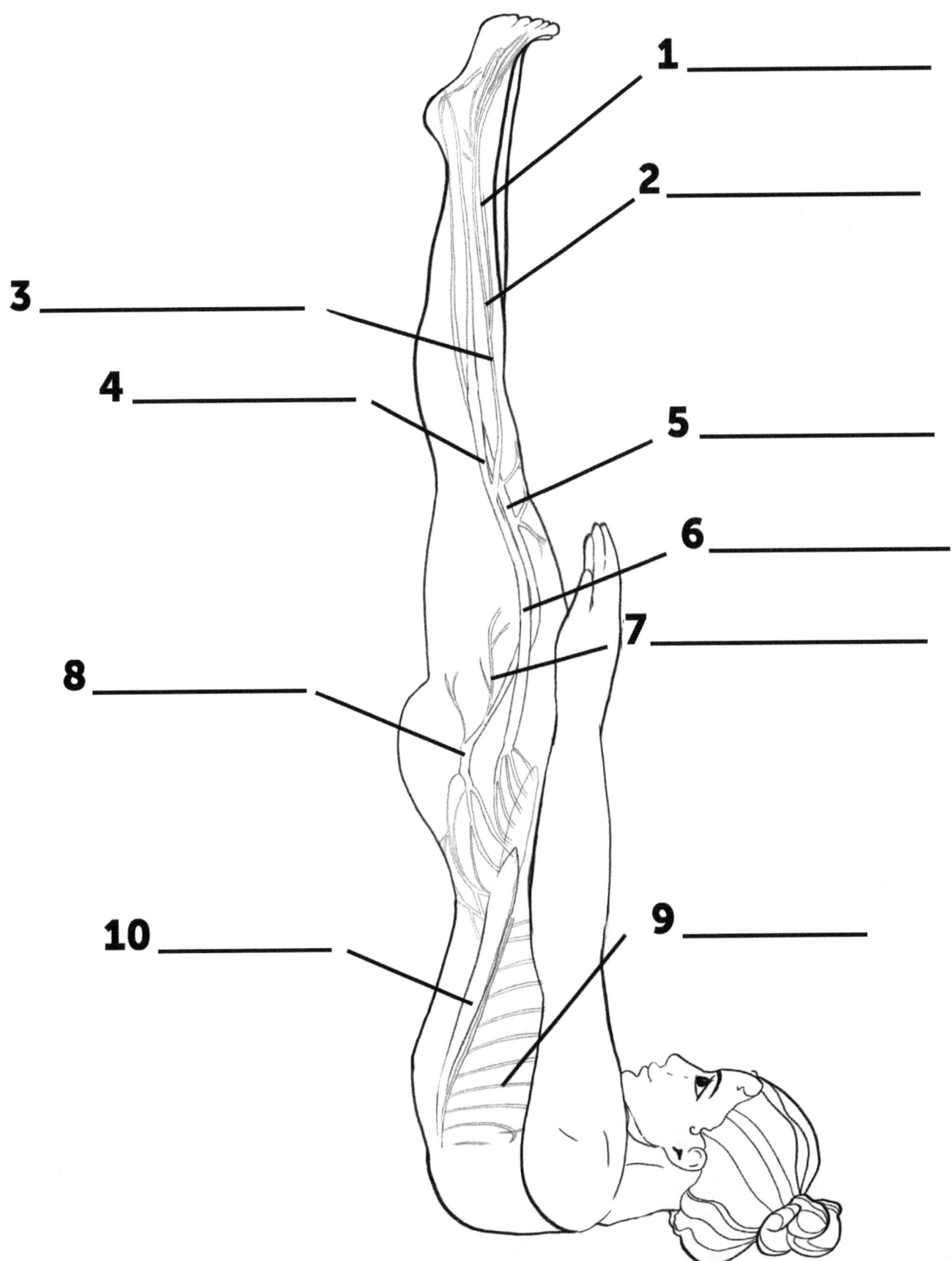

1 _____

2 _____

3 _____

4 _____

5 _____

6 _____

7 _____

8 _____

9 _____

10 _____

158. NIRALAMBA SARVANGASANA

1. PERONEO SUPERFICIAL
2. PERONEO PROFUNDO
3. PERONEO COMÚN
4. TIBIAL
5. SAFENA
6. CIÁTICO
7. RAMAS MUSCULARES DE FEMORAL
8. FEMORAL
9. INTERCOSTALES
10. MÉDULA ESPINAL

159. SKANDASANA

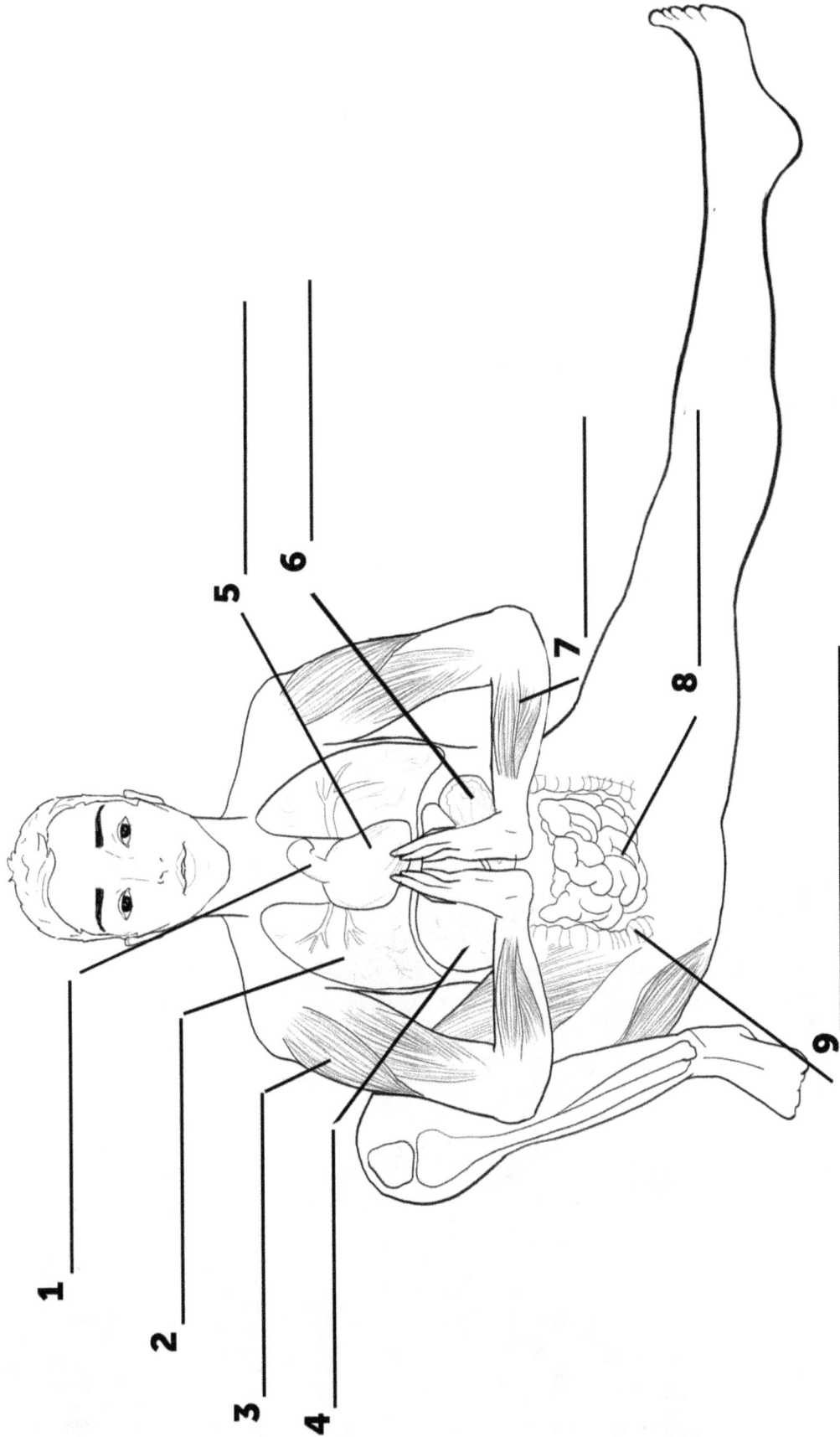

1

2

3

4

5

6

7

8

9

159. SKANDASANA

1. AORTA
2. PULMONES
3. DELTOIDES
4. HÍGADO
5. CORAZÓN
6. ESTÓMAGO
7. PRONADORES
8. FOLICULOS DE INTESTINO DELGADO
9. COLON ASCENDENTE

160. ANANTASANA PIERNA LEVANTAR

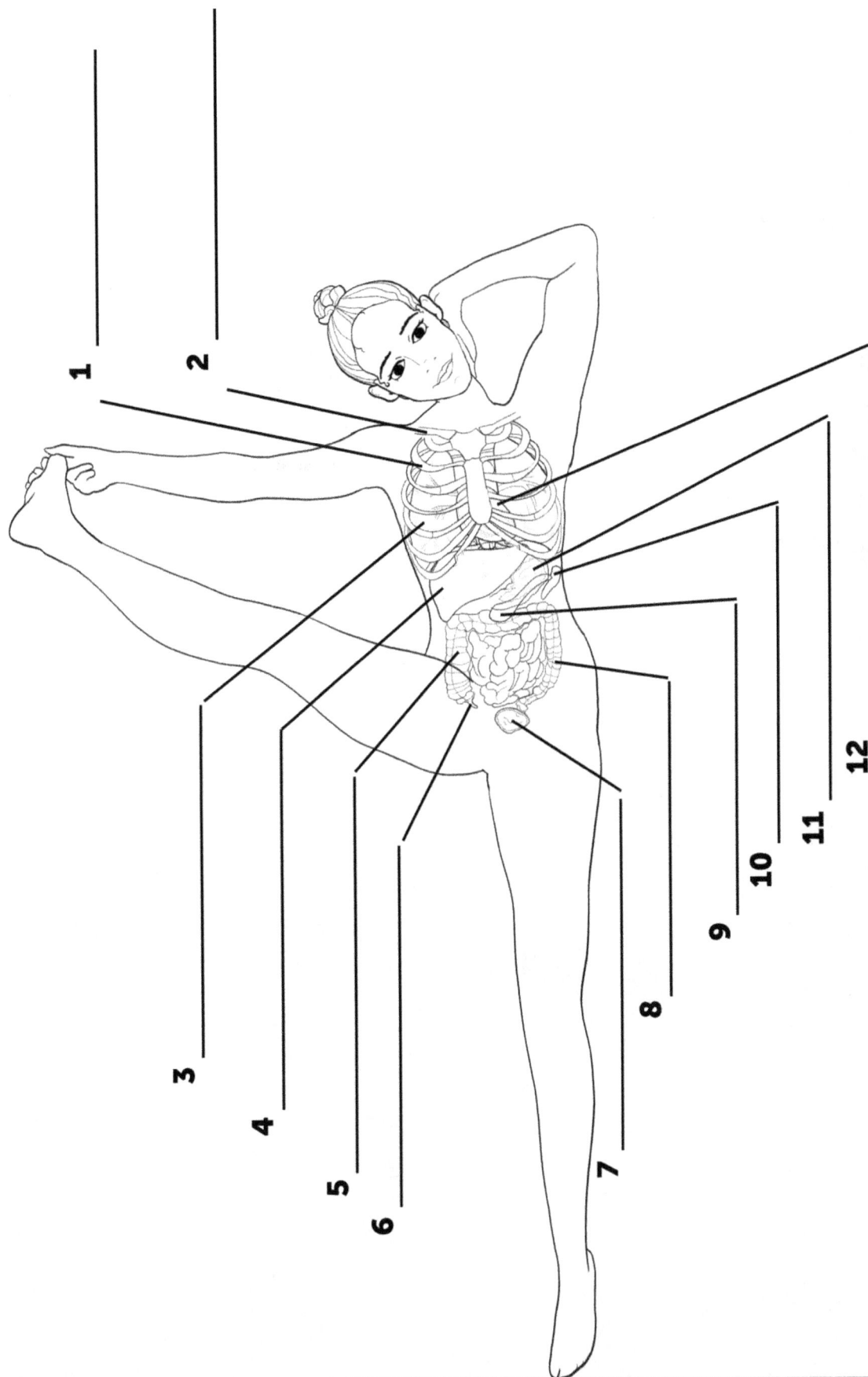

1

2

3

4

5

6

7

8

9

10

11

12

160. ANANTASANA PIERNA LEVANTAR

1. COSTILLAS
2. CLAVÍCULA
3. PULMONES
4. HÍGADO
5. COLON ASCENDENTE
6. APÉNDICE
7. VEJIGA URINARIA
8. COLON DESCENDENTE
9. PÁNCREAS
10. BAZO
11. ESTÓMAGO
12. CORAZÓN

www.ingramcontent.com/pod-product-compliance
Lightning Source LLC
Chambersburg PA
CBHW051204200326
41519CB00025B/7006